Outdoor Life

GUNS AND SHOOTING YEARBOOK

1987

Published by
Outdoor Life Books, New York

Distributed to the trade by
Stackpole Books, Harrisburg, Pennsylvania

COVER:

Jim Carmichel is pictured at the bench, preparing to test a Shilen DGA .22 CHeetah rifle with stainless barrel, Shilen fiberglass stock, and Bausch & Lomb 6–24× variable scope. Other guns shown on the bench, awaiting their test-shooting turns, are a Parker 28-gauge shotgun and a Remington XP–100 handgun in a new chambering—.223 Remington caliber with Leupold ER scope in B-Square mounts.

Copyright © 1986 by Times Mirror Magazines, Inc.

Published by
Outdoor Life Books
Times Mirror Magazines, Inc.
380 Madison Avenue
New York, NY 10017

Distributed to the trade by
Stackpole Books
Cameron and Kelker Streets
P.O. Box 1831
Harrisburg, PA 17105

Produced by Soderstrom Publishing Group Inc.
Book design: Jeff Fitschen
Editorial consultant: Robert Elman

ISBN 0–943822–84–X
ISSN 0889–0978

Manufactured in the United States of America

Contents

Preface *by Jim Carmichel*   v

Part 1:   RIFLES

A Custom Rifle: Can You Afford It?
by Jim Carmichel   2

Zero Check *by Col. Charles Askins*   6

Winchester Model 70: The Rifleman's
Rifle *by Garry James & Roger Renner*   10

The Ruger North Americans
by Ron Keysor   19

Good Shooting in Bad Light
by Bill McRae   21

Choosing a Pronghorn Rifle *by Bob Milek*   27

Part 2:   HANDGUNS

Trials of the .45 *by Col. Jim Crossman*   34

Hand Cannons for Big and Dangerous
Game *by J.D. Jones*   40

Mauser "Broomhandle": from the Boer
War to Today *by Garry James*   46

Return of the PPK *by Pete Dickey*   56

Part 3:   SHOTGUNS

Shotgun Fit: the Facts and the Fables
by Jim Carmichel   60

Shotgun Tips from the Masters
by Nick Sisley   64

Top-Performance Slugs and Slug Guns
by Brook Elliott   72

Inside a Patterning Circle *by Don Zutz*   78

**Part 4:   HISTORY AND
BLACK-POWDER SHOOTING**

Teddy's African Rifles
by A.D. Manchester   83

Black-Powder Shooters Square off Again
at 1,000 Yards *by Maj. Don Holmes*   87

Sir Charles Ross: His Controversial
Rifles and Cartridges *by H.V. Stent*   95

The Father of High Velocity
by Col. Charles Askins   101

A Man Named Weatherby
by Jim Carmichel   105

**Part 5:   AMMUNITION,
BALLISTICS, AND HANDLOADING**

Of Prime Importance *by Lon W. Sorensen*   110

Carmichel's Guide to
Hunting Cartridges *by Jim Carmichel*   118

Does the Wad Make a Difference?
by Hugh Birnbaum   123

Handloading Secrets for
Big-Game Hunting *by Bob Milek*   **128**

Reloading Steel Shot *by Don Zutz*   **134**

**Part 6:   IMPROVEMENTS,
ALTERATIONS, AND GUNSMITHING**

In Search of an Accurate Barrel
by C. E. Harris   **141**

Alvin Linden, Dean of Stockmakers
by Ludwig Olson   **150**

.45 Auto Innovations for Accuracy
by Jack Mitchell   **159**

What Ever Happened to
Aperture Sights? *by Karl Bosselmann*   **164**

Three Centuries of Sling Evolution
by Pete Dickey   **167**

Scope That Handgun *by Bob Milek*   **172**

Appendix: ANNUAL UPDATE

Gun Developments
by Jim Carmichel   **179**

Index   **184**

Preface

This is the third annual *Outdoor Life Guns and Shooting Yearbook*, which in the fickle world of book publishing makes it something of a success. Perhaps we're all the more successful because we're breaking most of today's book publishing rules. The first rule, I'm told, is to ride the crest of popular trends. But we follow no trends. In fact, I haven't even bothered to find out what they are.

We've ignored another rule by not basing the contents of this book on reader opinion surveys. When Bill Ruger was asked if he enlisted the help of any market surveys in designing his guns, he replied that anyone who doesn't know what a gun should look like has no business being in the gun business. That's how we feel about the firearm publishing business. We don't attempt to insult our reader's intelligence or strain his interest by second-guessing or by slanting the editorial content toward special interest groups.

You'll note that the table of contents is arranged according to general topic (rifles, shotguns, etc.), but that is only for your convenience. We have not divined that any special area of interest is "hot" this year and jumped overboard in that direction. The one fundamental guideline we've followed in selecting this year's articles is the same as before: that they be the best available.

The one presumption we've made, if it can be called that, is that a reader who likes guns and shooting will have wide-ranging interests that encompass all areas of firearms and shooting experiences. Thus we feel that even if your bag is gun collecting you'll still enjoy and appreciate a well-written and informative story on handloading. We don't want to leave anyone out. But most of all we want to give *you* the best of everything.

For example, "Zero Check" by the inimitable Charles Askins is not only easy reading but provides information that will help ensure the success of your big-game hunts this season and for all seasons to follow. Charlie's advice is timeless. If you are in love with shotguns, take a close look at "Inside a Patterning Circle" by Don Zutz and pick up a wealth of guidance in "Shotgun Tips from the Masters," in which Nick Sisley offers the wonderful wingshooting advice he has gleaned from some of the top shotgunners.

If you're a Teddy Roosevelt fan (and who isn't?) or just like African adventure, "Teddy's African Rifles" is irresistible and "A Man Named Weatherby" profiles the man behind the name that changed the face of modern guns.

Like to talk technical? "Of Prime Importance" gives you a leg up, and then some, on handloading and gives you an edge at your club range or out where the deer and antelope play. Jim Crossman's "Trials of the .45" is the best treatment ever given the beloved .45 Auto. You won't be an expert until you read this one. I could go on and on bragging about this year's yearbook because everything is so good or, like my computer-generated "Guide to Hunting Cartridges," provides reference material that puts you out front in clubhouse discussions and in the game fields.

As with our articles, we select the best photos and art available—so long as they clearly illustrate the subject. This year we have the most informative illustrations ever and a lot more color.

Of course in the collecting and editing of this, the *1987 Outdoor Life Guns and Shooting Yearbook*, we've had one great advantage. Guns and shooting are fascinating topics. When the best articles by the best writers are collected, the results are bound to provide many hours of exciting reading and the best possible references. We know you'll be turning these pages again and again.

Jim Carmichel

PART ONE

RIFLES

A Custom Rifle: Can You Afford It?

Jim Carmichel

Last year, during the annual convention of the Safari Club International at Las Vegas, a bejeweled and black-tied throng of big-game hunters looked on with considerable amazement and delight as the bidding for a custom rifle topped the $100,000 mark and went on to reach a record-setting $140,000. A few weeks later, at the Game-Coin Convention in San Antonio, Texas— another gathering of ridiculously rich sportsmen—the hallowed London firm of Purdey's exhibited a trio of bolt-action rifles bearing price tags that averaged around six grand. When I told their stiff-necked representative that better custom rifles were made in America for only a fraction of the price, he puffed up like a courting partridge and allowed that their price also included a telescopic sight. Seeing that I was coming close to creating an international incident, I bought a shooting coat and engraved whiskey flask to show that we were still friends. Linda, my wife, helped out by buying a Purdey shell bag that she figures will be great for carrying cosmetics on safari.

During the exchange, a distinguished-looking Englishman had been listening with obvious amusement. As we were leaving, he introduced himself as a 40-year employee of Purdey's and added, "Your chaps do make the best rifles."

I can't think of many hunters who are itching to shell out more than 100,000 bucks for a cased rifle such as the one auctioned at the Safari Club International Convention or are even eager to pay six grand for an "ordinary" Purdey rifle. Happily, fine custom rifles, even those by some of our best craftsmen, don't cost anywhere near that. As you already suppose, a fine custom rifle costs considerably more than an out-of-the-rack factory-made rifle. But is it worth it? What does a custom rifle cost? And, for that matter, what is a custom rifle? Oddly, many of the fine rifles built by today's top craftsmen aren't custom jobs at all, certainly not in the traditional sense.

Perhaps I'm being old-fashioned but, to my notion, a "custom" gun—or "bespoke," as English gunmakers and tailors say—is one that is shaped to fit the individual physique of the owner. A

This article first appeared in Hunting Guns

truly custom gun is not necessarily fancy or expensive. For example, if you have a rifle with a stock that is too long for best comfort or most accurate shooting, you might saw an inch or so from the butt. It then becomes a "custom" rifle because it has been specially fitted to your physical needs. Or, stated another way, your rifle now would fit you better than it does other shooters, just as your custom-tailored suit fits you better than anyone else.

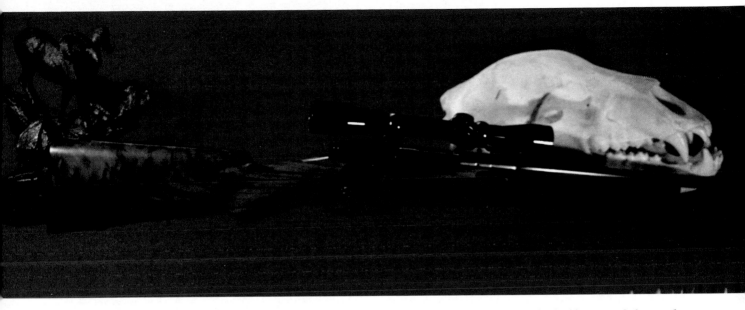

Pre-1964 Winchester Model 70 in 7mm Remington Magnum. The triggerguard is by Ted Blackburn and the stock, by Duane Wiebe, is California "English" walnut with 32-line-per-inch checkering. This beautiful custom rifle is equipped with 2–7× Burris scope in Conetrol rings. (Photos this page by John Sill)

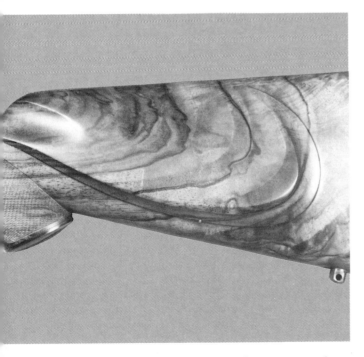

The cheekpiece on this Duane Wiebe custom stock exhibits extremely graceful shaping and beveling that serve to highlight and echo the wood's figure.

Above: skeletal pistol-grip cap framing checkered wood complements a fleur-de-lis checkering pattern which flows gracefully up the grip to the triggerguard. Note that the wood's handsomely figured grain shows through and blends with the checkering design. Below: The unique and intricate checkering pattern on fore-end is both esthetically and functionally perfect.

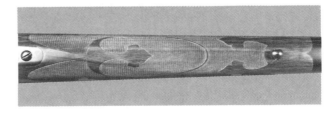

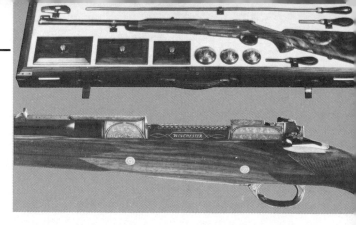

Built by David Miller Co., this cased custom .338 Winchester Magnum on a one-of-a-kind Model 70 Winchester action brought over $200,000 at auction.

Quite often, it takes more than shortening or lengthening the stock to make a rifle fit the owner most effectively. For instance, the comb may have to be raised or lowered or the shape of the pistol altered. Or, just as often, a hunter may want something different in the way of stock design just for the sake of being different or owning a one of a kind. This usually requires having a new stock made but, even so, the cost needn't be prohibitive.

Let's say you own a Model 700 Remington. It groups five shots in a half-inch circle at 100 yards, but you complain about the way it feels and you don't like the way it looks. Obviously, you don't want to get rid of such a wonderfully accurate rifle, but you would like for it to feel and look better. This calls for a "custom" stock, which can be acquired in any one of three basic ways. The first and least expensive of these is to simply make the custom stock yourself. You can get a stick of wood and whittle and chop a stock that is uniquely yours. Or, better yet, you can get a semifinished and inletted stock blank from a firm such as Reinhart Fajen. Most of the time-consuming work, such as inletting and rough shaping, is already completed with stocks of this type. All you have to do is final shape and finish to suit your requirements or whims. The price of a semifinished stock from Fajen runs from less than $50 on up to as much as you want to pay, depending on the type and quality of wood.

The next choice available in "custom" rifles is to have a firm such as Reinhart Fajen make and finish the stock. They are set up to build in all of the custom features that you can think of, at prices beginning at about $300. A third choice is to have the custom work done by your local gunsmith. Here, again, the work need not be expensive, depending, of course, on the custom features that you desire.

Back when I was in college and as poor as a one-flower funeral, I owned nothing but custom rifles. The reason being that they were all I could afford. I'd start off with a war-surplus Mauser or Enfield or Springfield rifle and replace the original barrel with a new one in whatever caliber I wanted. Then I'd fit a semifinished stock and, after some carving, sanding, and finishing, I'd have a pretty nice custom rig. Back then, the whole job would cost less than $100; today, it would cost twice that, but the price is still a bargain. There's an outfit in Connecticut by the name of Marathon Products that markets kits that include about everything you need to build your own "custom" rifle, including barreled action (rimfire and centerfire calibers), semifinished stock, and accessories.

Another way to get a "custom" rifle is simply to order it from one of the major gunmakers. Some of the big old-line companies still operate a custom shop that caters to customers' special needs. Most of the custom features that they install are simple touches such as extra stock length or changes in comb dimensions. They will also build you a stock out of extra-fancy wood or do some fancy checkering to special order. This work is usually quite expensive, sometimes costing more than it's worth, but they have to charge a bundle for their overhead and the usual bureaucratic paper-shuffling. But keep in mind that factory-original custom features have a tremendous collector appeal. After a few years, the cost of a custom feature done at the factory may have a collector's value of much more than what it originally cost. If you have some custom work done at a factory, be sure to keep the paperwork so you can prove that the work is "factory original." It's not unheard of for shady characters to doctor a rifle and claim that the modifications were done at the factory.

The most elegant—and certainly the most expensive—custom rifles of all are those that are built almost from scratch by one or more of the big-name craftsmen specializing in fancy custom work in wood or metal or both. Though the client (the difference between a customer and a client is about $1,000 or more) may specify a special dimension in the stock's length of pull or height of comb, more often than not these fancy rifles are built to the maker's standard dimensions. And, too, the client may have little or nothing to say about the overall shaping, finishing, and checkering—other than ordering the craftsman to do his best work. Some of the "custom" rifle builders send the client a price and check list to help him decide between, say, a skeleton buttplate and a recoil pad, and he may choose between checkering done in a point pattern or a fancy fleur-de-lis. Once these details are decided on, the craftsman pretty much does things his way and will tolerate little or no interference from the client. Some custom rifle builders don't even take orders, preferring to build the rifle to their personal tastes, then sell it on a speculation basis in much the same way that artists sell paintings or sculpture. Obviously, such a rifle can hardly be called a custom job because it wasn't built for anyone in particular. That's why I prefer to call them "art" rifles.

As you would expect, these art rifles can be mighty expensive because the maker almost always adds all the refinements that he can think of, plus having the rifle embellished by a top engraver and fitted in a fancy case. I've seen rifles of this type carrying a price tag of more than $30,000. But don't be misled by the asking prices of these extra-elaborate art rifles. After all, the maker is only fishing for someone who likes to throw his money around. To get an idea of what custom rifles are costing these days, I contacted Al Biesen, one of the world's best and certainly one of the best-known custom-rifle builders. Biesen does it all, fine metal work as well as stockmaking, and here's how prices run.

If you want a complete rifle with scope already mounted and ready to shoot, the price is $3,500. This includes a completely reworked Mauser action, a Douglas or Shilen barrel, a good grade of English walnut and, of course, Biesen's famous checkering and too many details to list here. If Biesen supplies a pre-1964 Model 70 Winchester action, the price goes up another $1,000, and you can add another $300 if you want a big magnum such as the .375 H&H. If you want a full-length Mannlicher-style stock, the final bill will be $400 to $500 higher.

But let's say you already have a rifle that you want customized by Biesen. For a re-stocking job, you'll pay $1,800 plus the cost of the wood and fittings. This basic labor charge includes total inletting and finishing and his best checkering. You can also get a skeleton buttplate and grip cap, if you want them.

Probably, you'll want some extra touches on the metal to complement the pretty stockwork, and here's what you can expect to pay: bolt-

Custom checkering is done by hand, so fancy patterns such as this one are very costly.

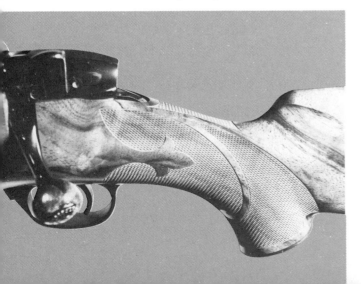

handle sculpturing, $75; bolt-knob checkering, $20 per panel; Model 70-style safety on Mauser action, $75; classic-style front sight ramp, $50; hood for ramp, $25; polishing action interior and jeweling bolt, $65; fitting, chambering, and turning barrel, $100; supply Douglas or Shilen barrel, $100; sculptured scope bases, $50 per pair; build new trigger-guard assembly (Model 70), $150; build new trigger-guard assembly (Mauser), $250; satin blueing (with complete job), $75; shorten Mauser action for short cartridges, $600.

If you want to go all out on a custom rig, you can think in terms of something like a David Miller rifle. The price here *starts* at $12,000. Engraving is extra. The philosophy behind a rifle of this type is doing everything possible to make it function as flawlessly as possible. For example, a metalsmith may spend up to *four* weeks just working on the action. Not building the action mind you, just improving it!

Can an incredibly refined rifle such as those built by the David Miller Company, or one of the gracefully elegant Al Biesen creations possibly be worth what they cost? Evidently, lots of folks think so because these and some other custom rifle makers have all the orders that they than can fill. If you think of a rifle as simply a tool for hunting, there is no way that you can justify the expense of a custom rifle. In fact, you'll probably be better off without one. Some hunters are intimidated by the prospect of hunting with such expensive equipment and would never be comfortable in the woods with one. Likewise, it would be difficult to defend the cost of a custom rifle in terms of performance. In my experience, custom rifles do not tend to be significantly more accurate than those that you might pick out of your dealer's rack. (Though I do have to say that my David Miller .338 Winchester Magnum, for whatever reasons, is the most dependably accurate rifle that I've ever owned. It *never* changes zero.)

In order for a custom rifle to be worth what you pay for it, you have to appreciate fine workmanship, or you have to be one of those tormented souls who won't accept anything but the best. If you love beautiful guns but also tend to be particular about getting your money's worth, the answer is still yes, a fine custom rifle is worth what it costs. Very often, they are worth more than what they cost because custom-rifle builders tend to be on the low end of the hourly wage scale. I recently had a bay window installed in my trophy room at a cost of more than $4,000. All I got was some glass, some ordinary wood and brick, and the hourly labors of a bricklayer and two carpenters. There is no way that that window equals either the refinement of materials or the wonderful craftsmanship incorporated in an Al Biesen rifle costing $3,500.

Zero Check

Col. Charles Askins

Proper sighting-in of 8mm–06 gave me good accuracy on African oryx hunt, despite harsh conditions.

For years, I hunted with a New Mexico cowboy who possessed one rifle, a Model 94 carbine in .30-30 caliber. This rifle stood in one corner of the saddle room until deer season popped up, whereupon old Buck slipped it in the scabbard under his left leg and rode off to stalk the wily mule deer.

"Don't you sorta get off a shot or two just to be sure that old fusee is still in zero?" I asked him one time.

"Nope," he dismissed without comment. I suspect the idea that the rifle would not shoot straight was beyond his comprehension. It had hit his buck last year so certainly it would be okay for the current season. The sights were the open type, about 14 inches apart, and the front bead had been busted off and replaced with a blob of welding metal.

Experience has shown me that old Buck's attitude about sighting-in is not uncommon. For two decades, I lived beside a public rifle/pistol range. It was a mile-and-a-half distant, and I could plainly hear the constant *rat-tat-tat* of gunfire. I was probably the most consistent user of that facility, and strange and wondrous were the feats of marksmanship I witnessed. I talked to a great many of the shooters, partly because I was interested in their case, but more, I suspect, because I wanted to get an idea of just how much background each marksman might have.

The range was a benchrest facility. The benches were sturdily built of two-inch planking set on 4×4 legs and must have weighed about 300 pounds. There was a movable block at the front end of the bench on which the shooter could pile his sandbags. The management was generous with the bags, which was a great help. The chair was a folding sort, often seen at concerts and similar activities, and was exceedingly poor. A sturdy bench needs a sturdy seat; we made do with the chair offered.

Most shooters had little idea of how to sit up to the bench. One reason, of course, was that there was no one to teach them. The old buzzard who managed the place had no intention of playing the part of coach. You could come out and bang away to your heart's content, and whether you ever really got the rifle in zero was of no concern to him. There were, quite naturally, good shots who took every advantage of the bench and sandbags, but these gunners were largely ignored by the rank and file.

One laddy-o confided to me, "I don't really need to waste all this expensive ammo on gittin' sighted-in. I wuz all right last year, so I know I'm okay now."

I asked whether he was shooting from the same box of cartridges.

"Nope," he admitted. "I went down to Nagel's Gun Shop an' bought me another box of shells."

He was certain that so long as the cartridges were the same caliber and bullet weight the new box would shoot precisely like the old. I explained to him that factory loads weren't that consistent, and unless he shot a few rounds from each box for zero, he could be in trouble. He looked at me like I had lost my marbles.

It was a majority problem that no shooting had been attempted since the last deer season. A common complaint was, "My wife knocked mah rifle off the shelf, an' I reckon it ain't shootin' right." Or, "My teenage son got it out of the closet and tinkered with the scope." These, I suspect, were quite legitimate bellyaches, reminding me of when

This article first appeared in American Hunter

the Army issued sniper rifles during World War II. The rifles, complete with the Weaver scope, were handed out to troopers who had utterly no experience with the scope-sighted ordnance. The Army had no sniper schools, and so no one knew anything about the scope-sighted rifle or how to use it in the sniping role. The admonition from the company commander was: "Now don't tamper with that telescope. It was sighted at the arsenal." This thinking was strictly poppycock.

There is, I observed, a surprisingly large number of shooters who still adhere to this philosophy. Their gunsmith mounts the scope on the newly purchased rifle, and the new owner be-

Arlen Chaney, President of Omark Industries' Sporting Division, checks zero of his .30–06 pre-'64 Model 70 in camp in Alaska's Wrangell Mountains before setting out to hunt Dall sheep. (Photos on this page are by Jim Carmichel)

Chaney here confirms that his rifle and cartridges are grouping where he wants them to group for Dall.

lieves that it is hitting dead center. The last thing he wants to do is to tinker with it.

At sometimes as much as a dollar per shot, you can scarcely blame some gunners for being a bit reluctant about burning up the costly fodder. Not everyone handloads his hunting ammo. Virtually unknown to the hundreds of shooters who passed through my public shooting grounds was the collimator. This handy device, when plugged into the muzzle of the rifle, presents a set of optical lenses that project a crosshair or a matted crosshatch. The marksman then maneuvers his scope to place the crosshairs in the collimator. This ensures a reasonably close hit on the target. The shot won't hit dead center, but it will be in the proximity. From this first-round hit, he can adjust his sight for elevation and deflection and fetch the rifle to a reasonable zero.

During the 20 years I shot on my range, I never saw a collimator, although they are readily available. Dean Alley offers a good one and so does Tasco.

It is pretty generally accepted that all serious hunters now use a scope sight. It was a never-ending source of amazement to me the number of shooters who turned up at the range, usually just before deer season, with iron-sighted muskets. One of these simple souls had a Model 95 lever-action with a 28-inch barrel. It was the .30-40 caliber. He sat down at the bench, and it was obvious he had never fired off the table before. He propped himself up on his elbows and placed the muzzle just ahead of the sandbags. When he fired, he promptly levered out the empty and whipped a fresh round into the chamber. This second shot followed the first in a period of not more than three seconds. "It's that second shot that gits 'em," he explained to me.

He kept on firing until he had gone through a box of new factory cartridges. In every case, he whipped in that second shot just as fast as he could work the lever. I looked down at the target through the spotting scope. This mark was a 30 × 30 backboard with a five-inch bull's-eye in the center. He had hit every corner, and one shot of the 20 had struck the bull.

Another character was pretty original about his practice. He sat down at the bench, but since the table was movable (even though it did weigh 300 pounds) he scooted it up against a 4 × 4-inch upright. He then leaned forward and rested the barrel, held by the left hand, against the post. He had a table filled with sandbags, but these he disdained. He liked that sturdy pillar. "It's jist like a tree in the woods. I always look fer me a tree," he explained.

He got off a couple of quick shots with the post as an aid, and then he leaped nimbly off his chair and plopped down on the ground in a sort of

After sighting-in from a bench, you should practice on a deer target from various ranges to test your accuracy under field conditions.

half-baked kneeling position. There he got off two more quick shots. Now, of course, this shooter wasn't trying to sight-in, he was simply practicing for the forthcoming deer season. At any rate, it was quite entertaining and showed originality.

After 40 years of hunting on all the continents save Australia, I never found in the game fields but two benchrests. One of these was on Olga Bay on Kodiak Island, and the second was on Terror Bay, on the other side of the same island. Not anywhere in Africa did I ever see any kind of rest that might have passed for a bench. I often speculated what went through the heads of the countless hunters with whom I shot. We'd go out to stir up such bad-tempered denizens as the lion, the leopard, or the Cape buffalo, and the professional who literally put his life in your hands had no more idea than Idi Amin whether you were sighted-in or not. It baffled me, I'll tell you.

I like a bench that is 30 inches in height and at least four feet in width. The depth is somewhat critical too, for the side of the table next to the gunner must be scooped out so that he can push his chest against it. The wooden bench is okay if it is quite massive. I have experimented with these portable jobs, and all are wobbly. The best is the bench cast of concrete. It is the most sturdy, the most secure, and generally the most acceptable. The surface should be carpeted to protect elbows and stock wood.

I have also tested adjustable muzzle rests, and these gizmos seem to be fairly substantial. However, I'd just a lot rather depend on a 6 × 6-inch block of wood to which I can add sandbags. That is, after all, the most secure.

The sandbags should be piled up in front and the forestock bedded in them. The barrel is *not* rested on any part of the bags. Behind, the sandbags are placed beneath the toe of the stock until the rifle is horizontal and pointed at the middle of the target without any maneuvering on the part of the marksman. When he sits down at the bench, there should be nothing left to do except grip the piece at the small of the stock and take aim.

There must not be any inclination to lift the toe of the stock off the sandbag. If it becomes necessary to left the rifle, even very slightly, then the gun is not properly set.

It is all a matter of adjusting the bags. If the crosshairs fall below the middle of the bull's-eye, then the front bags must be supplemented. If the aim is high, then the bag under the toe of the stock must be augmented. Meanwhile, the shooter wants to push up against the curved portion of the bench very strongly. The chest should bear across its whole front. The upper torso is none too stable, and unless an anchor is secured between the bench and the man's chest, he will never be sufficiently steady.

A great many benchresters simply do not know what to do with the left hand. They feel, somehow, that the hand ought to be on the fore-end beneath the stock and resting on the sandbags. Perish the thought. If the rifle is one of those light

A rifle's zero should be checked when a hunter reaches camp, even though it was sighted-in at home.

kickers, like the .270 or the 06, the left mitt plays no part. The rise of the muzzle will be so slight as to create no problem.

If the caliber is something over 8mm, however, I place a finger over the barrel ahead of the receiver ring. That dampens the tendency of the muzzle to rise under recoil—and smack you in the chops. The hand ahead of the receiver has no effect on the shot group. If there is a sling on the rifle, and most hunters have one, it plays no part in the exercise. It is better, as a matter of fact, removed for the benchrest stint.

Some contend the rifle will place its shot in one spot from the bench and in quite another when fired, say, offhand, sitting, or prone. I find this is hooey. If the shooter is careful to fire with the gun barrel free and ahead of the forward sandbags, he will find the zero is constant in the game fields.

It used to be we felt that a rifle was not properly tested unless a 10-shot group was banged off. Then there came along the school that elected to only five-shot groups. This was infinitely easier, and I can understand why a lot of fellows were none too enthusiastic about trying to group 10 rounds into a two-inch circle at 200 meters. Here more lately, most of us settle for a three-shot group. This may be the lazy man's way of sighting-in, but I like it.

In the first place, the game is very apt to be struck with the first shot out of a cold barrel. If the rifle has any tendency to do something awkward with that first round, it had better be discovered over the benchrest. I have seen any number of muskets that would group handsomely for the first three shots and then wander a bit for the next three-shot trial. After all, those first three shots are the critical ones, with lots of accent on Round Number One.

Personally, I always sight-in for 200 yards. This is a matter of choice, and the fellow who is going to hunt in the second-growth of Oregon's coastal woods may feel that 50 yards is quite sufficient. Each hunter is a law unto himself. He will contemplate his hunting area and decide how far his shot will be taken. This will govern his zero when he sights-in from the bench.

I always admire fellow firearm pundits who never fire anything less than one-inch groups at 100 yards. I think this is most commendable, but the mere telling of these one-hole exhibitions does harm to the poor fellow who only visits the range once yearly and then on the day before the deer season opens. He does well to hit the five-inch bull's-eye at least half the time. I'd speculate if the one-time-a-year venison chaser can group his first three shots into a three-inch circle at 100 yards, he is going to be a pretty lethal stalker of the elusive whitetail.

Winchester Model 70: the Rifleman's Rifle

Garry James and Roger Renner

It is a rare thing these days when a product actually lives up to the hyperbole of its advertising copy, but Winchester's Model 70, "The Rifleman's Rifle," has matched and perhaps surpassed the claims made for this fine arm.

Since its introduction in early 1937, this turn-bolt sporter has been chambered for some 31 rounds (if one wishes to include a possibly apocryphal .416 Rigby custom gun) and has been offered in a dizzying variety of styles and models. Its reputation has scarcely diminished.

Actually the 70 story begins in 1925 when Winchester decided to produce a high-grade, but affordable sporter, to introduce its new .270 round. The Model 54 came out initially in .270 and .30-06, though about eight rounds were eventually added to its repertoire. It was an immediate success, and soon outclassed all other American bolt-action game and target rifles. Like the Model 70 to come, the M54 was available in a number of styles aimed to suit just about any shooter's needs.

Despite the onset of the Depression in 1929 and the takeover of Winchester by Western Cartridge Company in 1934, the M54 remained fairly popular. Still, Winchester felt that a few improvements were warranted, and as early as 1934 work was begun to redesign this fledgling classic.

The Model 54 had racked up a few complaints and these were to be rectified in the new arm. Though lock time had been improved immeasurably, it still came up for criticism as shooters found the rifle was prone to occasional misfire.

Additionally, the M54's two-stage military-style trigger pull was not particularly conducive to match-grade accuracy, and the Mauser/Springfield-style flip-over safety was found to be awkward to use when the gun was fitted with a scope.

While the first "Model 70," as the new rifle was to be called, rolled off the line in mid-1936, the

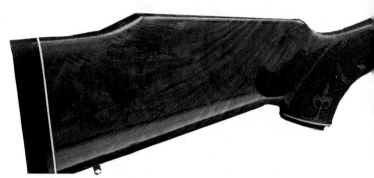

Collector Jerry Vallens purchased this special Model 70. Its serial number is 1,500,000 and it chambers the mighty .338 Winchester Magnum cartridge. The rifle is factory-engraved and both gold inlaid and overlaid with game animal motifs.

rifle was not actually announced until January of 1937, the year in which it went on sale.

Offered initially in five versions (the Standard Rifle, Super Grade, Target Rifle, National Match, and Bull Gun), though it harked back to the M-54, the Model 70 was really unlike anything yet seen on the American sporting market.

The Model 70 included such innovations as a more convenient horizontal safety, hinged floorplate, improved speed lock, new bolt stop, straighter stock design with 21-line-per-inch checkering, and forged steel triggerguard (as opposed to the M54's formed sheet-metal unit).

While the 70 was a more appealing and functional piece than its predecessor, the gun's real magic rested in its action.

Though some of the 70's "innovations" had actually been incorporated in late Model 54s, the newer gun was still seen by shooters as a considerable improvement. The two-lug Mauser-style

This article first appeared in Guns & Ammo

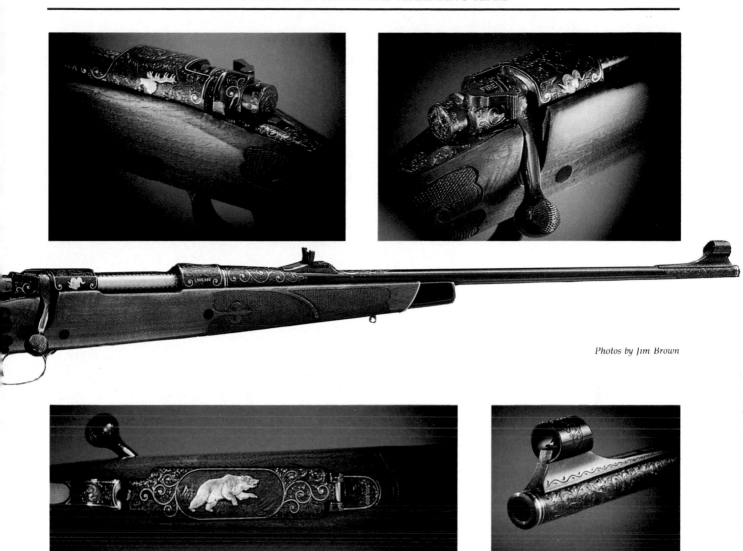

Photos by Jim Brown

bolt had a one-piece firing pin and was pierced with a pair of vent holes which, according to a 1938 Winchester brochure, "are there to relieve pressure in the firing pin hole in case of a ruptured primer." The gun was fitted with a typical Mauser-style long external extractor, which contributed to its controlled feeding.

The trigger system had been almost completely revamped, though again, it still had some features found on the 54. Without going into tedious mechanical details, let it suffice to say that the designers came up with a smooth, sure trigger that had a minimum of letoff. In fact, the basic device has hardly been changed over the past 38 years of the gun's production.

The Standard Model 70, in its 1937 guise, was a rifle that today would probably be considered a strictly custom item. The attention to detail on the metalwork, to include superb polish and blue, fit, and mating of wood-to-metal, was nigh onto impeccable. The good grade American walnut

stocks were available in low comb or Monte Carlo styles, checkered at wrist and fore-end. There was no fore-end tip, and the pistol grip was topped with a surprisingly tasteful sheet-metal cap debossed with the Winchester Emblem. The buttplate was of checkered steel. Sights, recalling the old days of the "factory-custom" Winchester lever guns, could be had in a variety of styles, including simple flip-up rear or receiver peeps.

Despite its name, the Standard Rifle was anything but standard. There were several different, widely varying calibers offered, as well as 20-, 24-, and 26-inch regular barrels, and 24- and 26-inch medium heavy tubes. As the years progressed, options increased accordingly, so an almost unheard of spectrum of choices was soon available.

The Super Grade, as its name implied, was, as per period advertising material, "built in general the same as the Standard Grade, with these refinements: Stock of selected walnut with or with-

WINCHESTER MODEL 70

Dates of Manufacture
1936 to 1981

Year	Serial Numbers
1936	1 — 2,238
1937	2,239 — 11,573
1938	11,574 — 17,844
1939	17,845 — 23,991
1940	23,992 — 31,675
1941	31,676 — 41,753
1942	41,754 — 49,206
1943	49,207 — 49,983
1944	49,984 — 49,997
1945	49,998 — 50,921
1946	50,922 — 58,382
1947	58,383 — 75,675
1948	75,676 — 101,680
1949	101,681 — 131,580
1950	131,581 — 173,150
1951	173,151 — 206,625
1952	206,626 — 238,820
1953	238,821 — 282,735
1954	282,736 — 323,530
1955	323,531 — 361,025
1956	361,026 — 393,595
1957	393,596 — 425,283
1958	425,284 — 440,792
1959	440,793 — 465,040
1960	465,041 — 504,257
1961	504,258 — 545,446
1962	545,447 — 565,592
1963	565,593 — 581,471
1964	700,000 — 757,180
1965	757,181 — 818,500
1966	818,501 — 855,860
1967	855,861 — 873,694
1968	873,695 — 929,990
1969	929,991 — 965,200
1970	965,201 — 1,008,436
1971	1,008,437 — 1,041,884
1972	1,041,885 — 1,088,291
1973	1,088,292 — 1,130,146
1974	1,130,147 — 1,176,878
1975	1,176,879 — 1,235,041
1976	1,235,042 — 1,298,272
1977	1,293,273 — 1,380,667
1978	1,380,668 — 1,423,869
1979	1,423,870 — 1,450,135
1980	1,450,136 — 1,493,463
1981	1,493,464 — 1,525,323

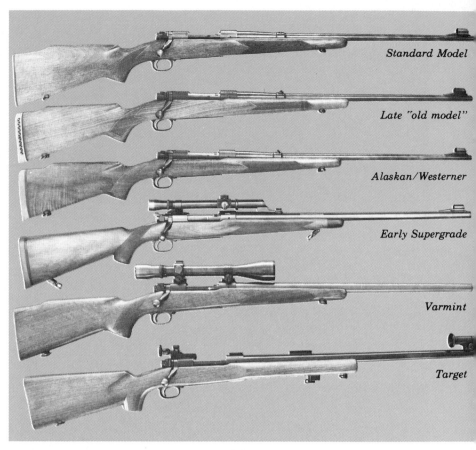

Standard Model

Late "old model"

Alaskan/Westerner

Early Supergrade

Varmint

Target

Above: The Pre-64 Model 70 was available in a number of different models with a dizzying spectrum of optional features. Below are just a few of the many post-64 variants of the famous Winchester Model 70 series.

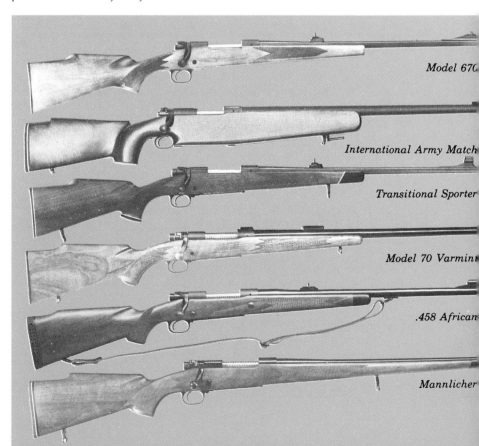

Model 670

International Army Match

Transitional Sporter

Model 70 Varmint

.458 African

Mannlicher

This postwar Model 70 Standard Rifle, with its Monte Carlo cheekpiece, is the classic American sporter. The .358 chambering (right) is scarce and now quite desirable.

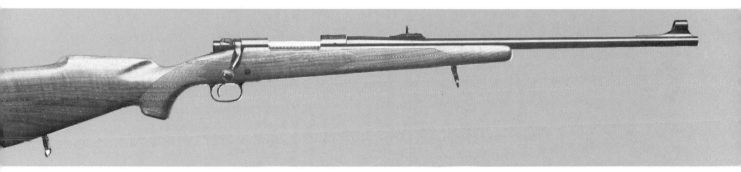

In the 1970s, Winchester announced the super-finished XTR series. This is a late 1970s Model 70A XTR.

out cheek rest, and with hard rubber grip cap, black molded forestock tip, and fine checkered grip and forestock." The gun also had detachable sling swivels, and a leather sling.

Other early Model 70 variants were the National Match with its target stock, free-floating barrel, and slightly more sophisticated sights; the Target Model, resembling the National Match in most details excepting its medium heavy barrel; and the Bull Gun, also maintaining the basic National Match look with the addition of a 28-inch extra heavy barrel.

As the years progressed, other models were added to the line such as the Featherweight, Varmint, African, Westerner/Alaskan, Mannlicher, and International Army Match.

The versatility of the arm was not limited to stock and barrel styles, however. The basic action was adaptable to a wide diversity of centerfire calibers ranging from .22 Hornet, through a healthy portion of the .30s, to the .458 Winchester Magnum.

Those Model 70s, made from 1937 to the beginning of World War II, are considered by many shooters and collectors to be the cream of American bolt-action sporting rifles. At the risk of sounding redundant, they exhibited an attention to detail, fit, and finish found only on some European production guns and a few Yank custom arms.

Even though everything from Spam to Donald Duck was pressed into service against the Axis, the Model 70, despite its admirable reputation for reliability and accuracy, was overlooked as a sniper arm in World War II. While it is rumored that some men took their personal Model 70s overseas to be used as special purpose arms, it wasn't until the Vietnam conflict that the sporter was finally drafted.

Though there are variants, the standard Model 70 sniper, as employed in Southeast Asia, was a heavy barrel .308 topped with an Unertl or Lyman target scope. In the hands of the Marines, these guns performed yeoman service.

Following World War II, the Model 70 appeared with much of its old pizazz, though signs of postwar austerity were starting to make themselves felt in the firearms industry. Checkering, while still hand cut, was reduced from 21 to 18 lines-per-inch, and there was some grousing from older sportsmen that wood, finish, and fit were not up to 1930s' standards.

To be sure, the Model 70 shot as well as ever and was comparatively still a thing of beauty, but some of the luster had dulled in the gray days of the Cold War.

Through the 1950s, along with this noticeable de-glamorizing trend, mechanization was conversely experiencing a considerable revolution. Technology was escalating in leaps and bounds and American industry could hardly keep pace with it. Economical production was the result, or more correctly, the target of this new revolution.

Many firearms makers applied this new technology as it became available, and several new firearms makers were born because of it. These "shortcuts" were applied to existing products gradually and weren't readily noticeable to the consumer.

Winchester, on the other hand, continued on in the traditional manner of gunmaking, which is to say they were losing money. The quality of the wood dropped and the hand checkering—which now was a custom feature in the trade—was becoming more coarse with each passing year. Finally, in the early '60s, cost effective measures became mandatory. A mechanical improvement was also in order and 1964 seemed as good a time as any to Winchester executives to make the change.

The resulting mechanical redesign wasn't actually as bad as many felt, but unfortunately, the "shortcuts," which Winchester had to now incorporate overnight, so soured the public that the new design improvements were viewed with the same contempt as the shortcut measures. It was the old "guilt by association" syndrome that suppressed the mechanical design.

Obvious differences, like the Parkerized receiver, alloy floorplate, and screw-on sights, turned the public off. But the real mortal sin was the light-colored stock, sporting darkened impressed checkering and a hogwallow for a barrel channel. These were just too much for the buying public to swallow at one time.

Taken individually, none of these features is necessarily bad. In fact, for the most part, many of the 1964 "goofs" are applied today by practically every production gunmaker. Winchester just gave us too much, too soon, thereby desecrating the rifle that by now had become an institution. Though the shooting public bemoaned the passing of prewar workmanship in the later pre-64s, and considered them inferior while in production, now they were beginning to look pretty good in comparison to the 1964 model!

Because so many pre-64s were still on the market, both new and used, no one had need to buy the new Model 70—and they didn't. This compounded Winchester's problem. Their own beloved Model 70 had become their worst competition.

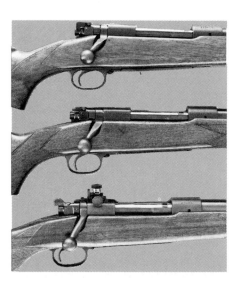

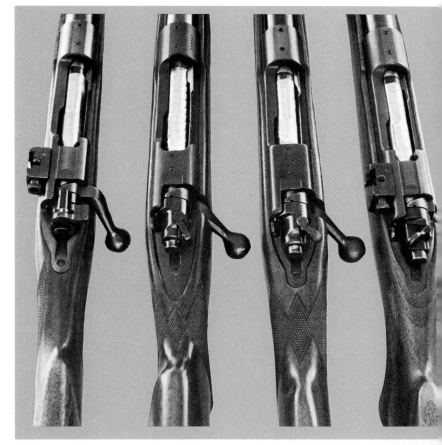

Above: The Model 70 safety evolved through three versions. They are (top to bottom) current style, transitional, early. Right: There has been quite a variety in Model 70 features. Most apparent are the "tangs" on the rear of the receivers and the checkering across the wrists of the stocks. The left-most rifle is the forerunner of the 70, the Model 54. It's easily distinguished by the turned-down bolt and "wing" safety.

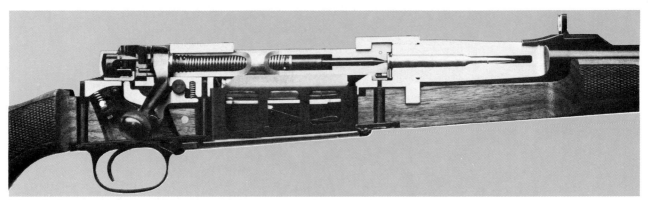

Though the cutaway Model 70 appears to be quite complex, the rifle's design is amazingly simple. It's one of few modern rifles whose bolt can be stripped and cleaned in the field.

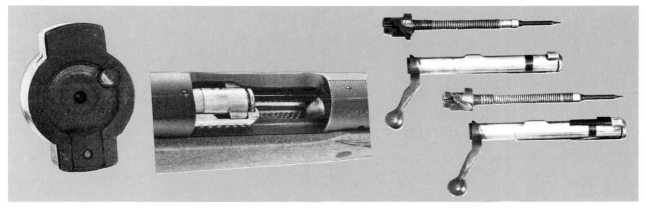

Comparison of the bolt face in the photo on the left to the bolt (middle) shows the small extractor in the locking lug of new guns, and the long claw extractor of the pre-64s. Photos at right show that the unusual extension of the locking lug on the lower bolt acts as a bolt stop in .22 Hornet Model 70.

The following years saw Winchester do a turnabout—albeit gradually. Their checkering was reversed with positive diamonds, the receiver's finish was given a bit more attention, barrel channels were reduced to reasonable widths, and all-in-all the guns became "acceptable." But it was too late. It seemed that Winchester would never live this one down.

Through the sixties another threat arose in the form of hordes of surplus military bolt-action rifles that hit the market. These were being "sporterized" by backyard gunsmiths everywhere, and sold for ridiculously low prices. A 1903 Springfield with a new Monte Carlo stock and fancy sights—or even a scope—could be had for less than a "C" note.

Again Winchester had to downgrade in order to compete for the shooting dollar. The Model 670 was born. It was a blonde-wood, blue-metalled baby offered in three versions—Carbine, Standard, and Magnum. Deleted was the three-position safety and the floorplate. Instead we were given a two-position side-tang safety and a blind magazine. Later a slightly jazzier version, termed the Model 770, was offered with a few more refinements. Through it all, the Model 70 proper remained pretty much the same with subtle upgrading in the form of finer finish, tighter inletting, and more artistic checkering patterns. Unbeknownst to the public, Winchester was not sitting home nursing its black eye. Instead they sat quietly back in the corner and flexed their new technological gunmaking muscles.

Later the 670s and 770s were dropped from the line and a few new ones, such as the Mannlicher and International Army Match, made their appearance in 1968 and 1970, respectively. In 1972 Winchester again offered a slightly downgraded Model 70A as we entered the recession of the 1970s. Things remained quiet in Winchester's corner through the 1970s, but they kept flexing those muscles.

Then in 1981 all hell broke loose in the gun business. Winchester came on the market with a vengeance, announcing the reintroduction of the Model 70 Featherweight after an 18-year absence. The new rifle was absolutely gorgeous and the world was agog with wonderment!

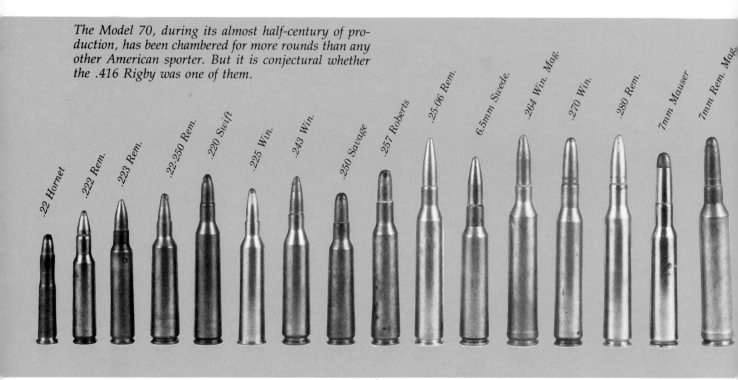

The Model 70, during its almost half-century of production, has been chambered for more rounds than any other American sporter. But it is conjectural whether the .416 Rigby was one of them.

.22 Hornet .222 Rem. .223 Rem. .22/250 Rem. .220 Swift .225 Win. .243 Win. .250 Savage .257 Roberts .25-06 Rem. 6.5mm Suede. .264 Win. Mag. .270 Win. .280 Rem. 7mm Mauser 7mm Rem. Mag.

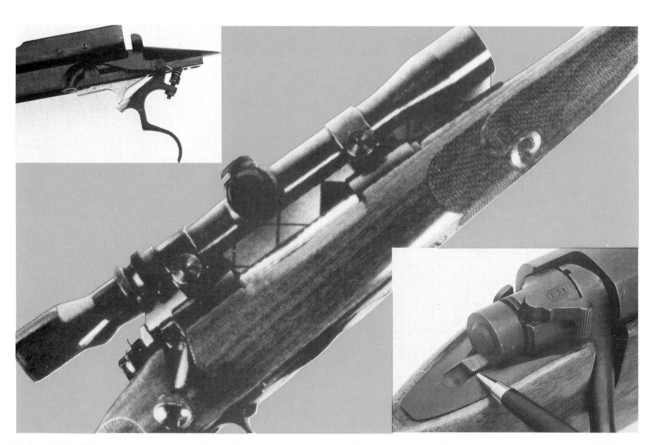

Today's Model 70 is a tribute to the craftsmanship and legendary accuracy of its predecessor. Inset left: Through the years the Model 70's simple and well designed trigger has remained unchanged. Inset right: The cocking indicator marks post-'64 70s; however, the early three-position safety was retained.

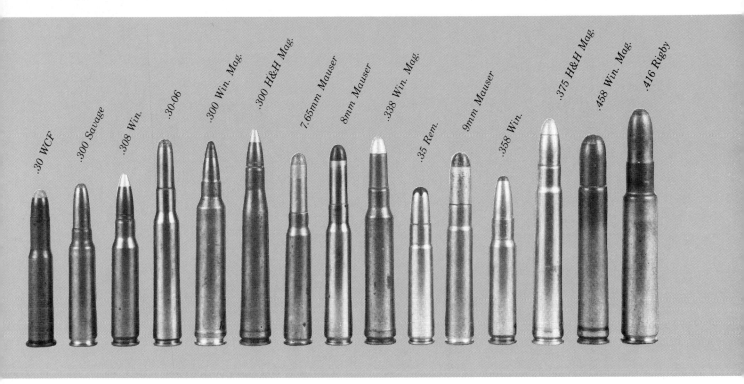

Gone was the clubby stock, the afterthought sights, the sandblasted finish, the angular checkering patterns, white-line spacers, and space-age plastic finish that everyone remembered from the recent past.

Instead, here was a slim, trim rifle sporting a classic stock with Schnabel fore-end, satin finished high-grade walnut, perfectly executed, borderless, cut checkering in the most pleasing patterns, and metal finishing that rivaled many of the Super Grade Model 70s of the early days. Today's Featherweight should wear the "Super Grade" logo on its floorplate—it has surely earned the right.

Shortly after the introduction of the Featherweight, Winchester (actually Olin) announced that its entire line of firearms would be made under license by a new company—United States Repeating Arms (USRAC). This further shook up the industry and the shooting public. Many thought that this meant that Winchesters would now be made in Japan as some other brands were. Actually, nothing has changed except the name, as it did when Olin Mathieson Chemical took over Winchester back in the 1930s.

The factory is right where it always has been and virtually the same people are making the same guns here in the U.S.A.

However, under the leadership of USRAC, the new Model 70 has regained the status of its predecessor. Winchester has been redeemed. No longer should they bear the stigma of "1964."

There is a new sense of direction at Winchester. They now have their ear in the marketplace and are in tune with giving the shooting public what it wants—quality at a reasonable price. For the first time it is possible to have a short-action Model 70 for those short rounds. Carbine and varmint variants, as well as the Standard and Super Express models, are also available in some new calibers, such as .280 and .25-06 Remington, and some old favorites like the .257 Roberts and the 7 × 57 have made guest appearances in recent 70s.

Starting in September 1985, U.S. Repeating Arms opened the doors of its Model 21 Shop to the custom Model 70 market. Now, as in the old days, you can order a Model 70 to suit your taste. Have your dealer contact USRAC for details.

Today's Model 70 is the result of 60 years of development, beginning in 1925 with the Model 54. Combine this experience with the precision of today's computer-operated machinery and a dedication at U.S. Repeating Arms to recapture the Winchester mystique, and the result is a modern classic.

The authors wish to thank the following for their help in the preparation of this article: Michael Kokin, Pony Express Sport Shop, Sepulveda, California, Wallace Beinfeld, Jerry Vallens, and Tom Rawson.

For those who would like to delve further into the history and development of the Model 70, we recommend *The Rifleman's Rifle*, by Roger C. Rule, Alliance Books, Inc., 18714 Parthenia St., Northridge, CA 91324.

Above: Artist Gary R. Swanson's butting bighorn rams painting is rendered in gold by engraver Franz Marktl on the first Ruger North American. The grip cap ram is from another Swanson painting. Right: The Rocky Mountain bighorn sheep rifle in .30–06 travels in its oak and leather Marvin Huey case. The rifle was sold in 1985 by Christie's auction house.

The Ruger North Americans

Ron Keysor

The quarter rib, which mates with the Leupold scope's Kimber mounts, and the operating lever are engraved and inlaid. Bill Dowtin did the stockwork.

Aside from the pre-1964 Winchester Model 70, with its legendary mystique, no modern American rifle has served better as a candidate for customizing and adornment than the Ruger No. 1. It took decades for the Model 70 to assume its classic status, hastened by its redesign in 1964. The Ruger No. 1 was an instant classic with its introduction in late 1966 and has served as a basis for innumerable custom rifles both here and in Europe.

Thus the appearance of another magnificent customized No. 1 rifle in itself is not a particularly noteworthy event. However, when the rifle bears the imprimatur of the manufacturer, Sturm, Ruger & Co., Inc., and of William B. Ruger, himself, it takes on added panache. And when it's the first

Engraver Franz Marktl's artistry decorates the front sight by Mark Penrod. Each rifle in the Ruger series will be distinctively customized.

of 21 rifles, each priced at $45,000, and each customized by some of the top American gunmaking artisans, the story builds.

The .30-06 illustrated here, the first of what have been dubbed the "Ruger 21 North Americans," is unique among its brethren. The rifle, donated by Ruger, was sold by Christie's at an unreserved auction in New York City in 1985. The proceeds from its sale will go toward a projected $3 million expansion of the arms and armor galleries of the Metropolitan Museum of Art.

Though coincidental, the sale of the first of the Ruger North Americans to benefit the prestigious Metropolitan intertwines with another unusual aspect of the series, a collaboration with some of the nation's leading wildlife artists. Each rifle in the series is dedicated to a North American game animal, and two original paintings by each of the artists commissioned to participate will be rendered in gold inlays on each of the rifles by master engraver Franz Marktl of Phoenix.

The first rifle, the .30-06, honors the Rocky Mountain bighorn sheep, and the paintings from which Marktl rendered the inlays were from the brush of Arizona artist Gary Swanson. Among others with commissions for the rifle series are Greg Beecham, Tom Beecham, Lee Cable, Bob

This article first appeared in American Rifleman

Kuhn, and Leon Parson. The purchaser of each rifle will receive a signed and remarqued color print of the painting used to adorn one side of the rifle's receiver.

Though Bill Ruger credits Dave Wolfe of Wolfe Publishing Co., Prescott, Arizona, with the concept of the Ruger North Americans series, Wolfe says the three-year, 21-rifle program evolved from conversations with engraver Marktl. Wolfe, publisher of *The Rifle* and *Handloader* magazines, is responsible for selecting the artists and artisans for the series.

While additional gunmakers will participate in the rifle construction, the .30–06 bighorn sheep rifle's barrel was made by Jim Baiar of Columbia Falls, Montana, and fitted and chambered by Ross Billingsly of Dayton, Wyoming. Billingsly also made the quarter rib with an express sight with one standing blade and one folding leaf, the band ramp front sight, front band swivel, and custom safety button. Additional metal work, including installation of a steel trigger, surface grinding the action flats, and fitting a new hinge pin was performed by Mark Penrod of Penrod Precision, North Manchester, Indiana. The stock and checkering are by Bill Dowtin of Flagstaff, Arizona. The rifle comes with a fitted leather and oak case by Marvin Huey of Kansas City, Missouri, with handmade accessories by Mike Marsh of Sheffield, England. The French-fitted case is in turn protected by a zippered green canvas case with leather appointments. Engraving is by Franz Marktl.

Though it's unlikely this rifle ever will be fired, these craftsmen have created an arm that handles as nicely as it looks. Weighing just 6 pounds, 11 ounces, sans its fitted Leupold Compact 6× scope in Lenard Brownell-designed, Kimber-made lever mounts, the rifle has a 22-inch barrel .562 inch at the muzzle.

The highly figured California English walnut stock and semibeavertail forearm by Dowtin carry 28-line-per-inch checkering with difficult-to-execute fleur-de-lis ribbons inside the checkering pattern. The butt is checkered, as well. A raised panel for the rear sling-swivel base, a knife edge atop the forward comb, and a finely cut shadow line around the cheekpiece are nice touches.

Penrod performed the action customizing. This included making an enlarged, checkered safety button and a finely sculpted trigger. The action tangs have been tapered, the factory's operating lever hinge pin hole reamed and a new pin and screw fitted, action surfaces refinished, the quarter rib detail completed, and standing and leaf sights added. Other Penrod touches include the handsome banded ramp front sight with gold bead, the rear swivel base, and the sling-swivel barrel band.

Marktl's engraving, of course, sets the rifle apart from all others. Beginning with the gold-inlaid bighorn sheep in panels on both sides of the gray-finished receiver, bordered in heavy bands of inlaid gold, it also carries the sheep motif on the gold-inlaid steel grip cap. "Ruger North Americans" is inlaid in gold on the barrel, as is the caliber marking, with "Limited Edition" spelled out in gold on top the quarter rib. The well-known Ruger eagle symbol in gold adorns the operating lever. Gold inlay, often finished with geometrical patterns, is everywhere, from the front sight ramp to the quarter rib to the sides of the operating lever, setting off the abundance of Marktl's scroll engraving which even extends to the scope rings. Marktl tastefully manages this extravagance of gold and engraving. The oversize forearm retaining screw's head (with its engraved escutcheon) and action pin and screw heads are decorated with rosettes, the slots, of course, precisely aligned. The rifle is signed by the engraver and marked 1 of 21.

"We wanted to show the world what really can be done with this rifle," said J. Thompson Ruger of the Ruger North Americans series. As vice president, marketing, for Sturm, Ruger & Co., Inc., he is managing the project for the firm. Each of the rifles in the series, he explained, will be distinctively customized, with calibers, barrel weights, and other features selected to be appropriate for the animal portrayed. Each rifle will be accompanied by a certificate of authenticity signed by William B. Ruger, co-founder, president, and chairman of the company.

The full list of 21 animals includes the bighorn, Dall, Stone, and desert sheep, mountain goat, grizzly, brown, polar, and black bears, bison, caribou, elk, Alaskan moose, mountain lion, jaguar, mule and whitetail deer, pronghorn antelope, javelina, wolf, and wolverine. The wolverine, his outsized reputation as a brawler notwithstanding, may find his inclusion here puzzling.

A number of the 21 rifles are currently under construction or completed, and they will be making appearances at events such as Safari Club International's convention, Game Coin, The Foundation for North American Sheep convention, and the SHOT Show.

Persons interested in purchasing rifles in the series may contact J. Thompson Ruger at Sturm, Ruger & Co., Inc., Lacey Place, Southport, CT 06490.

Limited-edition prints, 950 full-color lithographed copies of the painting that accompanies each rifle, signed by the artist, will be sold. The prints, costing $95 each, are being marketed by M/W Enterprises, Inc., 12629 No. Tatum Blvd., Suite 164, Phoenix, AZ 85032. Each will be serially numbered, with serial number preference considered for those ordering a full series.

Good Shooting in Bad Light

Bill McRae

It was 5:31 P.M. and there were exactly 10 precious minutes of legal shooting time left. The sun had disappeared behind an escarpment to the west an hour and a half earlier—it had officially been down for only 20 minutes—and heavy, snow-promising clouds had moved in from the north, cutting out the little sky light that remained. It would be dark night in the mountains and, with camp two miles away through choice grizzly country, I had an uneasy feeling that I should be going. It was already so dark that objects in the bottom of the draw at timber's edge were indistinguishable but, being a die-hard, I'd have one last look around.

The meadow (a park to Westerners) was located on a steep mountainside. Tracks and droppings indicated that bull elk were feeding there nightly and, by staying until dark, I had hoped to see them. A battered old Bausch & Lomb 7 × 50 binocular gave a picture, amazingly bright and clear, as it opened the shadows of the park's perimeter. There, halfway up the other side and about 100 yards away, two bulls were feeding.

Shouldering the Model 70 Winchester .30-06. I full well intended to send a 180-grain-Nosler greeting to the larger of the bulls but, when I looked through the 2.5 × scope mounted on the rifle, my heart sank. Instead of the bright picture that I had gotten through the binocular, I could see only faint light blobs that I assumed were the bulls. Worse still, the scope's thin crosshairs were not visible at all. A shot still might have been possible, but I deemed it too risky. Following wounded game with a flashlight doesn't appeal to me and, with snow in the offing, my blood trail would be obliterated by morning. Besides, elk, if left unattended, spoil quickly.

This article first appeared in Hunting Guns

Failing to·get the elk was a great disappointment for a young hunter, but there were valuable lessons in that long-ago experience. I already knew that the edge of night, either morning or evening, was the best time to hunt. But I learned the importance of a good, low-light binocular. The real revelation, however, was that a riflescope should be equal in light-gathering ability to your binocular.

While good light-gathering optics are most useful at dawn and dusk, under certain conditions they can be helpful any time of day. For example, thick tamarack or cedar swamps can be mighty dark on heavily overcast days. Even on sunny

GLOSSARY OF OPTICAL TERMS

Magnification or power: Let's use the 7×50 binocular as an example. The first figure, 7×, denotes the magnification. This means that distant objects will appear seven times larger than when viewed with the naked eye, or conversely, you will view them as though they were one-seventh of their actual distance away.

Objective Lens: The objective (front) lens is the lens nearest a viewed object when the binocular is in use. The number immediately following the "×" in the identification number (such as 50 in 7×50) is the objective lens diameter expressed in millimeters.

Ocular lens or eyepiece: The ocular lens is the lens that is held nearest the eye when the binocular is in use.

Exit pupil: Hold the binocular at arm's length and point it at a broad source of light. The small circle of light that you see in the eyepiece is the exit pupil. Exit pupil diameter is stated in millimeters and is computed by dividing the diameter of the objective lens by the magnification. A 7×35 binocular, for example, has a 5mm exit pupil (35 ÷ 7 = 5).

Relative brightness: This factor serves as a traditional guide for comparing image brightness between binoculars of different magnification and objective lens sizes. To determine relative brightness, divide the diameter of the objective lens by the magnification and square the result (which is the diameter of the exit pupil). For example, an 8×40 binocular has a 5mm exit pupil and, thus, a relative brightness of 25. Even under the lowest light conditions, the human eye cannot utilize a relative brightness greater than 50.

Twilight factor: A newer and perhaps more reliable guide for comparing low-light performance of binoculars is the twilight factor. It differs from relative brightness in that it takes magnification into account. To determine the twilight factor, multiply the magnification of the binocular by the diameter of its objective lens, then find the square root of the result. For a 10×40 binocular, the formula looks like this: 10x40 = 400, and the square root of 400 is 20. Thus, the twilight factor for the 10×40 binocular is 20. The higher the twilight factor, the more detail will be visible in dim light.

Eye relief: This term refers to the distance that the ocular lens (eyepiece) of the binocular must be held from the user's eye in order for the binocular's full field of view to be visible. Long eye relief is important for those who wear eyeglasses.

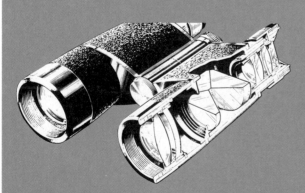
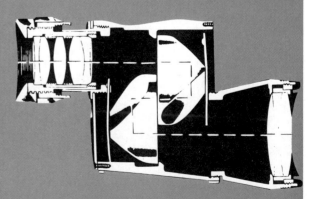

Left is a cutaway view of a Bushnell roof prism binocular. It clearly shows the relationship between: (1) the objective lens, which gathers an image in the form of light, (2) the prism, which transmits the image, and (3) the ocular lens, which projects the image to the viewer's eye. Right: This cutaway is of a porro prism binocular. Everything said about the roof prism binocular also applies. The only difference is in the design.

days, because your eyes become accustomed to the bright light, it's hard to see into deeply shaded spots, which are the places where game is most apt to bed. A bright scope or binocular will open up the shadows and let you see what's there.

What makes binoculars or riflescopes suitable for low-light use? There are three major factors: exit pupil diameter, which translates directly to relative brightness; magnification, a term interchangeable with power; and light transmission. (As you read on, you may want to occasionally refer to the "Glossary of Optical Terms" on the previous page.)

EXIT PUPIL SIZE

The exit pupil is a bundle of light rays that emerges from the eyepiece of an optical instrument and enters the viewer's eye when it is positioned correctly at the proper eye relief. It can be seen as a round spot of light in the eyepiece when the viewing instrument is held at arm's length and pointed toward the sky or toward a brightly lit background. The amount of light that the exit pupil can transmit is determined by its diameter. To find the metric diameter of the exit pupil, divide the diameter of the objective lens by the rated magnification. Thus, the diameter of the exit pupil on a 7×35 binocular is 5mm $(35 \div 7 = 5)$; on a 4×28 riflescope, the diameter is 7mm $(28 \div 4 = 7)$. Relative brightness is the square of the diameter of the exit pupil and, using the preceding examples, it is 25 and 49, respectively.

Why bother to square the exit pupil's diameter? Because by doing so, we get its area as opposed to its diameter, and area better indicates the exit pupil's ability to transmit light. Note that, when the diameter of the exit pupil is doubled, its *area*

is quadrupled and four times as much light is transmitted. For example, if you compare a 7×50 binocular (exit pupil diameter is 7.14 and area is 50.98) with a 7×25 binocular (exit pupil diameter is 3.57 and area is 12.745), the 7×50 transmits four times as much light as the 7×25. Other things being equal, the 7×50 weighs roughly four times as much, while the diameter of its objective lens doubles. Compared to a 7×35 (exit pupil diameter is 5 and area is 25), a 7×50 (exit pupil diameter is 7.14 and area is 50.98) transmits twice as much light, weighs roughly twice as much, and is about 30 percent larger.

The diameter of the objective lens (the lens nearest a viewed object) controls brightness in a very simple way. The larger the opening, the greater the amount of light that goes into the optical system—it's like comparing large windows to small ones.

From this, you might conclude that, to make scopes or binoculars brighter, all a manufacturer need do is increase the objective lens diameter. This is true, but only to a point. The first problem, as we've already seen, is that the size and the weight of an instrument increase with the increases in objective lens diameter. However, the immutable, limiting factor is the entrance pupil of the human eye, which plays a role in the optical system of any visual instrument. The eye has an iris diaphragm that determines the size of the eye's entrance pupil according to the prevailing light conditions. The pupil gets larger in low light, smaller in bright. The catch is that, even under the lowest light conditions, the pupil of the human eye can open up only to 7mm and is seldom larger than 5mm in most low-light hunting situations. If an optical system provides an exit pupil with a diameter larger than that of the pupil of the human eye, the added light is wasted.

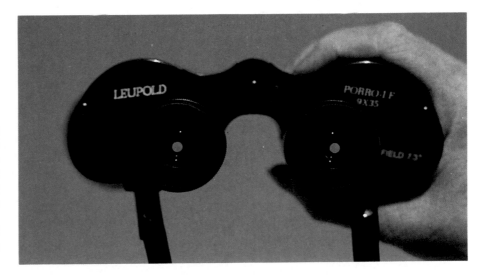

The round spots of light in the center of this Leupold 9×35 binocular's eyepieces are the exit pupils. Exit pupil size is a major factor in determining low-light performance—the larger the exit pupil the better. The author prefers a 7mm exit pupil in a binocular and an 8mm exit pupil in a riflescope.

It should be noted, however, that an instrument with a slightly oversized exit pupil is more comfortable to use because it allows more latitude in the position of the user's eye. The hunter does not have to place his eyeball exactly in line with the exit pupil.

MAGNIFICATION

Magnification significantly improves low-light performance, and the twilight-factor formula takes this into account. The way this formula works is simple. Magnification makes objects appear larger, and the larger something appears—especially in low light—the easier it is to see.

Let's suppose that, on a wilderness hunting trip, you left your horse tied to a tree and, as a result of hunting too long, you must find the horse in the dark or walk several miles back to camp. In the daylight, it had been possible to see the horse a mile away but, in the darkness, you must get within 10 yards before you can see it. Now, for the sake of simplicity, let's say that you are using a 7× binocular that delivers an image equal in brightness to the naked eye (most 7× glasses do much better). You will now be able to see the horse at 70 yards because, through the binocular, it will appear as large as if it were only 10 yards away. You could see the horse at 80 yards with an 8× binocular, at 100 yards with a 10×, and so on.

It is important to note here that, while increased magnification does make objects appear larger and thus easier to see in poor light, it does not make them brighter. This is where the twilight factor, if carried to the extreme, breaks down. High magnification is helpful *only* if it is combined with a generous-sized exit pupil. A case in point is a 30×60 spotting scope: It has a very high twilight factor of 42 (see the "Glossary of Optical Terms" to learn how this was computed), but has an exit pupil diameter of only 2mm and a relative brightness of four. This makes it nearly useless in low light.

LIGHT TRANSMISSION

The third major factor in low-light performance is light transmission. It is the very important but little-talked-about stepchild in the world of optics. Relative brightness and the twilight factor are mathematical formulas that only tell how instruments *should* perform in low light. Theoretically, binoculars or scopes of equal magnification power and with lenses of the same objective diameter should be equally bright. In the real world, however, optical instruments that have the same

identification numbers do not necessarily perform equally well under low-light conditions—and the reason for this is the light-transmission factor.

Unfortunately, there is no industry-wide standard for evaluating and reporting light transmission—in spite of the fact that light transmission can range from as low as 40 percent in systems with uncoated optics to more than 90 percent (70 percent is good).

Among the things that can affect light transmission, antireflection coating is the most important. It reduces the amount of light reflected away from the highly polished glass surfaces on the lenses of a binocular or a spotting scope. The coating material most often used is magnesium fluoride. An optimum thickness of this, deposited on an optical surface, will reduce the amount of light lost from that surface from about 5 percent to about 1 percent. When you realize that a binocular or riflescope may have 10 or more air-to-glass surfaces, it is easy to see why antireflection coatings are important. The new multicoatings are even more effective.

Other factors affecting light transmission are the clarity of the optical glass and the detail-robbing glare that is most noticeable when you're looking into the sun at dawn or dusk.

Without uniform standards, how can you be sure that the optics that you select yield good light transmission? The best assurance is to buy from a manufacturer that has a reputation for quality and dependability. Unfortunately, this rule of thumb makes short shrift of some small manufacturers who produce top-quality instruments.

Another option is to visually compare different models. Look into deep shadows to see how bright they are and then toward a light source to check for glare. You'll be surprised at the differences that you find. As Ken Woytek, riflescope engineering manager for Bausch & Lomb points out, "The human eye is not of itself a good instrument for judging optical quality, but it is great for making comparisons."

BINOCULARS

What does it take to have a good low-light binocular? At the risk of being disagreed with or even renounced, I am going to arbitrarily set minimum standards for both relative brightness and the twilight factor. Here they are. Relative brightness should be at least 25, with a twilight factor of at least 17—and not one or the other, *both*! This leaves out all of the compacts and many fine full-sized binoculars. By this standard, the venerable 7×35 doesn't make it—its twilight factor of 15.6 is too low. The much-proclaimed 10×40 doesn't

Although quality light-gathering optics will do yeoman service at dawn and dusk, you may find them most useful when peering into deeply shaded spots, which are the places where game is most apt to bed down. A bright scope will help open up shadows.

cut the mustard, either—while it has a high twilight factor of 20, its 4mm-diameter exit pupil yields a relative brightness of only 16. I said that I was being arbitrary!

Realizing that I am ignoring other binoculars that are equally deserving, I am going to talk about some of my favorites. The 7×50 Bausch & Lomb is one and, because it has been around for a long time, it has fought in several wars and has sailed the seven seas. Its almost-identical twin, the 10×50, is also a fine low-light glass.

A Bausch & Lomb binocular that really excites me is the new 8×42 Roof Prism Elite. This glass is the nurtured child of optical engineer Al Akin. In fact, Al was so fussy about having his personal quality standards met on this binocular that production was delayed for more than a year. His goal was to make "the world's best binocular." Whether he succeeded or not, I'm not qualified to say but, when looking through the Elite, the claim isn't hard to believe. It has wonderfully long eye-relief, is very bright, very sharp, easy on the eyes, and priced accordingly.

Zeiss binoculars have impressed me to the point that my wife describes our family's financial status as "Zeiss poor." Every Zeiss binocular is good—you can count on it—but some are better than others, particularly in low light. With a relative brightness of 49 and a twilight factor of 21.16, the Zeiss 8×56 is probably the king of the night glasses. A lighter and smaller Zeiss, almost its equal, is the 7×42. It has a generous 450-foot field of view and, at 28 ounces, it is a great all-around hunting glass.

Remember, these are simply examples of fine optics. There are others, indeed.

TELESCOPIC SIGHTS

What whould you look for in a riflescope? Beginning with the exit pupil, a scope that is going to be used in low-light conditions should have an exit pupil no smaller than 7mm in diameter—8mm is even better. Why 7mm instead of 5mm, as was the case with binoculars? Because scopes have exceptionally long eye-relief, which makes perfect eye alignment harder to achieve, and the larger-diameter exit pupil leaves more room for lateral eye movement. Incidentally, the $2.5 \times$ scope

This is a cutaway view of a Bushnell riflescope. A 3×–9× variable like this one is a good choice because low magnification can be used in good light at close range or for fast action. But when the light level gets low, the magnification can be turned up to increase the scope's twilight factor.

that let me down on that elk hunt wasn't lacking in exit pupil size. With its 20mm objective lens, it had an 8mm exit pupil, but it lacked magnification, good light transmission, and a decent reticle.

High magnification helps but, on a fixed-power scope, it presents a problem. It enhances low-light performance, but it is a detriment when shooting in thick cover or at running game. The solution is a variable power scope, preferably with a 3×-to-9× range. Such scopes usually have the large objectives needed to gather light, and the twilight factor improves as the power is turned up.

In low light, especially in the evening when there is little time for tracking wounded game, pinpoint shot placement is a must. It is vital that the reticle be seen clearly. Among the conventional reticles available to American sportsmen, the Duplex, which was invented by Leupold and copied by everyone else, is the best.

Bushnell's Lite-Site riflescope is another option. Under normal light conditions, the reticle of the Lite-Site looks like any other dual-thickness crosshair but, in low light, a switch is turned on that causes a red dot to be projected in the center of the reticle. The 3×-to-9× model that I tested gave a bright image, and the dot was very apparent, even in situations where any other reticle would have been lost. However, be sure to check the hunting regulations for the area where you'll be shooting before using this scope. As of last season, it was illegal to use the Lite-Site in Pennsylvania. Other states may have similar regulations.

Good low-light performers that grace rifles in my gun cabinet are: Leupold's Vari-× III 2.5×-to-8×, a Bausch & Lomb 3×-to-9×, and a Zeiss 4×. The latter two have multicoated lenses, which may be why they have never let me down.

As far as recent developments go, Bruce Cavey, sports optics division manager for Zeiss assures me that, by the time you read this, a Zeiss 8×56 riflescope will be available in this country. The 8×56 has long been a favorite in Europe where hunting is often done in low light. I have ordered one, mainly because I'm so fond of my 8×56 Zeiss binocular. It will go on a flat-shooting Remington Model 700 .25-06 and do double duty on varmints. The price will be about $510.

Another point of interest is that Nikon, a company that is world renowned for fine cameras, is making a strong entry into U.S. sporting optics. Along with a very fine line of binoculars, they are offering two multicoated riflescopes, a 4× and a 3×-to-9× variable. Both have 40mm objective lenses, and the optics are what you'd expect from Nikon—bright and sharp!

Now some loose ends. What about spotting scopes? For reasons already noted (small exit pupils), they are not very useful in low light.

Do yellow shooting glasses help in low light? Definitely not. Yellow glasses filter out scattered blue light, thus increasing contrast and sharpening detail on foggy or hazy days. They are, however, optical filters, and all filters reduce—in one way or another—the total amount of light that passes through them. We want to increase light, not reduce it.

All optical instruments perform best if they are kept clean. I've seen hunter's scopes and binoculars that were so dirty I could hardly see through them in *good* light. I could only imagine how useless they would be in poor light.

And, finally, three very important points. Know the legal shooting hours and follow the law to the letter. Play it safe—if you don't know where the bullet is going, don't shoot. And don't shoot if you're not sure of making a clean kill. You owe this to yourself, to the sport, and to the game.

Choosing a Pronghorn Rifle

Bob Milek

The pronghorn antelope is a resident of America's Western and Southwestern plains, a land so vast and empty that it takes away the breath of someone accustomed to hunting in confining hardwood and evergreen tangles. In this country, they say you can look into tomorrow from any hill or ridge. If the hunter can see into tomorrow, the pronghorn, with his spectacular eyesight, must see clear into next week. Many a time I've located pronghorns several miles away and found them looking right down the tube of my spotting scope.

As a pronghorn hunter, the first thing to bear in mind, then, is that if you can see a pronghorn buck, chances are good that he can see you. The second thing to remember is that once he does see you, he takes off like the wind and in a matter of seconds can put a half mile between him and your rifle. In other words, you want to move carefully and take no chances on spooking your quarry. Many times you'll find it better to try a long shot at an undisturbed buck than to attempt to work closer in country so devoid of cover that a rattlesnake can't hide.

Right off the bat this should give you a hint as to what you're looking for in a rifle/cartridge combination for pronghorn hunting. The rifle should be scoped, capable of at least 1½ minute-of-angle accuracy, and should develop enough velocity that the bullet can be delivered on target at 400 yards, maybe more, without a lot of guessing. Likewise, the remaining velocity of the bullet should be great enough at 400 yards to adequately penetrate and expand.

This sounds like a tall order, but in reality you have any number of cartridges to choose from. Of course, you can eliminate most of the low-velocity cartridges from any list considered for pronghorns. Old favorites, like the .30-30, .32 Special, and .35 Remington, superb cartridges out to medium range, don't cut the mustard in open country. But how do you decide where to draw the line in cartridge selection for pronghorn hunting?

I guess the place to start is to learn just a little about the physiology of the pronghorn. This prai-

This article first appeared in Guns & Ammo

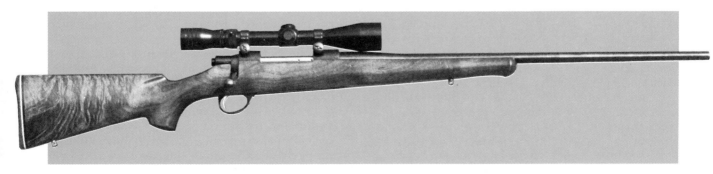

Milek's combination of this custom-stocked .25–06 Sako and Redfield's superb illuminator optics is typical of the kind of shootin' machine that will take pronghorns a quarter-mile away!

Because the bulk of pronghorn hunting is spent glassing wide expanses of the plains, a quality rifle sling is an essential piece of equipment which allows the hunter use of both hands to steady the binocular. Because the rifle spends so much time on the shoulder, a wide, comfortable sling is important.

rie speedster is actually a small animal, much smaller than he appears. A mature buck will stand about 30 inches high at the shoulder, weigh between 120 and 140 pounds, and the vital area of the chest where you must place the bullet measures no more than 13 inches top to bottom. A pronghorn has thin skin and a small bone structure.

Penetration, even when you have to break the shoulder, is not a problem. However, the pronghorn can be one of the most tenacious of North American game animals when wounded. If one gets away you'll have a devil of a time tracking him over dry ground that's usually laced with pronghorn tracks. You have to place your shot carefully and put some of the vital organs out of commission. With either a heart or lung shot, a pronghorn will take off running the moment he's hit, but he's dead on his feet and will go down within 50 or 75 yards.

I'm a great believer in stalking pronghorns, working hard to get just as close as possible before shooting. Usually you'll be able to get to within 200 yards or less of your quarry when you do things right but as I mentioned earlier, there are times when the terrain precludes anything except taking a long shot. It would be hell to have to pass up your one chance at a trophy pronghorn simply because your rifle and cartridge weren't capable of getting the job done at 400 yards or so.

The correct approach, then, is to use a cartridge that is capable of a one-shot kill at 400 yards. After a lifetime of pronghorn hunting, both professionally and for my own sport, I've come to the conclusion that a cartridge for pronghorn hunting should deliver the bullet on target with a minimum remaining velocity of 2,000 fps—about what's needed in any caliber to afford proper penetration and reasonable bullet expansion on an animal the size of a pronghorn. In other words, if you're going to be trying 400-yard shots, you'll want to use a cartridge whose bullet will arrive on target traveling at 2,000 fps or more. If you think you're good enough to place a shot accurately at 500 yards, then that bullet had better get there at a speed of 2,000 fps.

Now before you panic, remember that this is only a guideline. However, it does place such cartridges as the .243 Winchester, 6mm Remington, .250-3000 Savage, .257 Roberts, 7mm-08 Remington, 7 × 57 Mauser, and .308 Winchester right on the borderline. I'd be the last person in the world to suggest that these cartridges be dropped from the running where pronghorn cartridges are concerned. I've killed a number of pronghorns with all of them. However, I am saying that if you use one of these cartridges, select your bullets and loads carefully and do everything you can to keep your shots under 400 yards.

Undoubtedly the best cartridges for pronghorn hunting are those which shoot bullets weighing at least 120 grains at muzzle velocities of 3,000 fps or more. In standard cartridges this means the .25-06, the .270 Winchester, the .280 Remington, and the .30-06. In magnums there's quite a list; the .264 Winchester Magnum, 7mm Remington Magnum, .300 H&H Magnum, .300 Winchester Magnum, and the Weatherby line up through the .300 Weatherby. While the .338 Winchester and .340 Weatherby are superb long-range catridges, there just isn't any earthly reason for such big bores for pronghorn hunting. In fact, there are many shooters who'll argue that there's no place for any magnum larger than the 7mm Remington in pronghorn hunting. True, you may not need the power of the .30 caliber magnums, but if you can shoot them well, they're excellent long-range numbers.

My own taste is for the nonmagnum cartridges and among these the .25-06 is my favorite for pronghorns. I use spitzer bullets weighing 117 to 120 grains and push them out the muzzle at around 3,100 fps. These shoot a little flatter than a 100-grain spitzer even though the latter leaves the muzzle faster. Then, too, I've found that the 100-grain .257-inch bullets are a bit thin-skinned for .25-06 velocities when shots at 200 yards or so are involved. Too often the 100-grain designs will go to pieces at normal ranges, doing a lot of unnecessary meat damage.

This brings us to the matter of bullet selection, an extremely important subject for pronghorn hunting. Because pronghorns are small animals, hunters too often tend toward light bullets in any caliber. This is the wrong approach, particularly where high velocity and magnum cartridges are concerned. Sure, you can send those lightweights a-whizzin', but stop and think about two things. First, what's going to happen when that lightweight, thin-jacketed bullet hits a little pronghorn at 150 or 200 yards at high velocity? I'll tell you what happens—it goes to pieces! Sure, because a pronghorn is small you'll get penetration, but you'll also have bloodshot meat from the neck to the rump, rendering a lot of it inedible. Shooters of .30 caliber magnums are particularly guilty of this mistake. They'll go for a 150-grain bullet that, designed for use at .308 and .30-06 velocities, will blow a pronghorn darn near in two at 150 or 200 yards. However, a 180-grain bullet, with its heavier jacket, will expand enough to do quick-killing organ and tissue damage, yet it will hold together, most likely exiting the body, and do a

According to Milek, the lineup of cartridges at right would be sufficient for taking pronghorn out to about 300 yards. For longer shots, more potent medicine is necessary.

minimum amount of damage to the meat. However, don't go to the real heavyweights in the magnums because they have jackets that are so tough that on a pronghorn, even at close range, expansion will be minimal or nonexistent and the bullet will punch right through like a pencil, doing little or no damage to the vital organs.

Second, lightweight bullets in any caliber may start out considerably faster than heavy ones, but because they have poorer ballistic form, or coefficients, they'll shed their velocity faster. This means that their trajectory won't be as flat as that of the heavier bullets. When you get out around that 400-yard mark, the heavier bullet, which started slower, will be traveling almost as fast as the lighter design. Of course, where long-range open-country hunting is concerned, a spitzer bullet design is the only choice. The spitzer bucks the wind better than any other and it overcomes the friction of air better, so it reaches a far-off target with optimum velocity and accuracy.

Once you've decided on your cartridge for pronghorn hunting, it's time to address the matter of the rifle itself. Considering the list of cartridges that are suitable, it's obvious that all action types enter the picture—single shots, lever actions, pumps, semiautomatics and bolt actions. Which will you choose? Begin by looking at the accuracy requirements of the rifle—1½ minute of angle, or approximately 1½ inches at 100 yards.

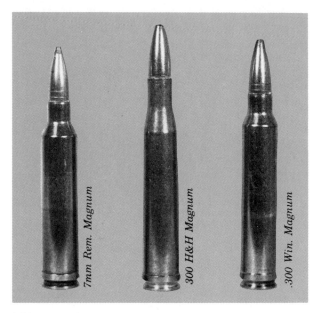

This cartridge lineup represents the upper limits of practical calibers for taking pronghorn.

Author's son took this buck with a Remington 700 BDL in .280 Remington, which is a fine combination.

Everything being perfect, such a rifle will group its shots in about six inches at 400 yards. This really isn't too stringent an accuracy requirement, but you'd be surprised how many sporting rifles won't shoot this well right out of the box. Often you'll find it necessary to rebed the barrel and action and work the trigger over to get such accuracy from a rifle, and in some cases, everything you try won't produce 1½-inch groups.

Your best chance of getting the required accuracy from a sporter will be with a bolt-action rifle. The bolt is the strongest, most rigid action available. Equally important is the fact that the bolt rifle uses a one-piece stock that bolts solidly to the barreled action. The action and recoil lug can be glass bedded and the barrel can be free-floated, bedded tight, or free-floated with a pressure point up near the tip of the fore-end. All of these are ways you can work with a bolt-action stock to improve the accuracy of your rifle. With no other action design do you have so many options that can influence accuracy. Then, too, bolt-action rifles are usually fit with triggers that can be adjusted or honed to produce a very good trigger pull that's a big help in shooting accurately. If the trigger isn't adjustable, it can sometimes be replaced with a custom trigger.

Next to the bolt-action rifle, the single shot is probably your best bet. While there's usually not a lot that you can do with the bedding of the action and barrel on a single shot—the stocks are of two-piece design—single-shot actions are strong and the rifles have fine triggers. Likewise, single-shot rifles are chambered for a number of

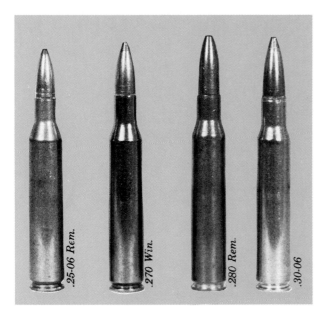

These four cartridges are based on the .30–06 and represent the optimum for pronghorn hunting at ranges out to 400 yards. Milek prefers .25–06 Remington.

excellent cartridges for pronghorn hunting. There are shooters who criticize single-shot rifles because you have only one shot. I don't see this as a detriment on a pronghorn rifle. In fact, it may be helpful. When you know you only have one shot, with no quick follow-up, you're more inclined to take your time and make that first shot count.

Among the remaining action types—lever, slide, and semiautomatic—there's not a whole lot of difference in how they perform. Those that are chambered for good pronghorn cartridges are strong, but in most instances they have poor triggers that can't be adjusted or replaced and there's no stock work you can do to make them shoot better. Still, there are lever, pump, and semiautomatic centerfire rifles that shoot very well, often meeting that 1½ MOA criterion. The problem is that if you buy such a rifle and it won't shoot up to snuff, there's just nothing you can do about it.

Weight is a concern in selecting a pronghorn rifle, but not as much as it might be for a rifle to be used for mule deer and elk hunting in rough canyon and mountain country. Even when you have to do a lot of walking after pronghorns, you won't find the going difficult. As long as your rifle doesn't weigh over 8½ or 9 pounds, you'll probably get along okay. However, the lightweight rifle craze is having an effect on pronghorn rifles just as it is on rifles for other big-game hunting. There's just something about a light rifle that attracts hunters. But with few exceptions, the lightweight designs available to-

day are built on short actions and these can't handle the cartridges that are best for the long-range work encountered in pronghorn country. Also, most of the high velocity pronghorn cartridges, particularly the magnums, need a barrel at least 24 inches long in order to develop the optimum velocity. Long barrels just don't go hand in hand with what we've come to regard as lightweight rifles.

As I mentioned earlier, a good scope sight is a must on a pronghorn rifle. There's no way that you can shoot accurately at long range with open sights. But what magnification is best? Should you be looking at a fixed-power or a variable-power scope? There was a time when a 4× fixed-power scope was the standard for big-game hunting and no one ever questioned the wisdom of this. Not so today. In open country, 6× is a better choice, offering just enough extra magnification to be helpful, yet not so much that it magnifies your normal wiggles to the point that they become unnerving.

Likewise, where fixed-power scopes were once the only thing for hunting, variable-power scopes are now pushing them for popularity. This is due in part to the fact that the variables of today are far better scopes than those we had around 10 or 15 years ago. Optically they're on a par with the best of the fixed-power designs and they provide a degree of versatility that many shooters find quite attractive. If a variable-power scope is your choice for a pronghorn rifle, one of 3-9× is probably your best bet with 2-7× running a close second. Simmons has a new 4-10× wide-angle variable-power scope with an adjustable objective that may be excellent on a pronghorn rifle. Gen-

Another of the author's rifles is this Mark X Mauser wrapped in custom wood from Bishop, topped with Leupold optics. The caliber, .25–06, is just about optimum for pronghorn hunting.

For years the fixed 4× scope, like the Leupold (top photo) was considered standard for plains shooting, but today's hunters are turning to variables, such as the excellent Redfield Illuminator 3–9×, for extra versatility.

Another of the author's favorite antelope rifles is Browning's new A-bolt wearing a Simmons 4-10× scope and chambered in—you guessed it—.25–06. Bench shooting the new A-bolt turned in the groups at left, which are great for plains hunting.

Though weight isn't a major concern in pronghorn rifles, lightweights, like the Ultra Light Arms .308 at 5½ pounds complete, are appealing!

Single-shot rifles like the T/CR '83 are good choices for pronghorn hunting as they are usually chambered in calibers perfectly suited for the job and are strong and accurate for long shots.

erally speaking, variable-power scopes are bulkier and weigh a little more than fixed-power designs. These are the only disadvantages I see in them.

We've discussed cartridges, actions, and scope sights for pronghorn hunting, so I guess it's time to look at what I personally consider the ideal pronghorn rig. Would you believe it—I have four pronghorn rifles, all different, but each works fine for hunting pronghorns in the open plains coun-

try here in Wyoming. All four have two things in common—they're all bolt-action rifles and they're all chambered for the .25-06 Remington cartridge. Two of these are custom rifles, and two are production rifles. One custom rifle uses a Sako barreled action set in a classic stock and topped with a Redfield 3-9× Illuminator variable-power scope. The other is a Bishop Model 10 rifle which has a Mark X barreled action in a beautiful piece of American walnut crafted by Henry Pohl of Bishop Stocks. This one carries a 4× Leupold scope set in Conetrol mounts. For production pronghorn rifles I have an ancient Model 700 BDL Remington and a Browning BBR that I've just topped with the new Simmons 4-10× scope. Naturally, my pronghorn rifles are fitted with slings, making them handy not only for packing the rifles in the field, but for steadying my shooting positions.

You may not agree with my choices of pronghorn rifles, but that's fine. After all, a hunting rifle is a personal thing that reflects individual tastes. But agree or not, you'll do well to heed what's been discussed concerning cartridges, bullets, sights, and accuracy. I've spent a lot of years hunting the speedy pronghorn and I know what it takes to come home time after time with a buck. During my years as a professional guide, some of my clients chose to ignore my advice about rifles and cartridges for pronghorns and more often than not their decision came back to haunt them. Remember, pronghorns live in open country, they have superb eyesight, and they're the fastest animals on the North American continent. Combine these three things and you have need of an accurate, flat-shooting rifle.

PART TWO

HANDGUNS

Trials of the .45

Col. Jim Crossman

Those who think the recent "selection" of a 9mm double-action pistol as America's official Personal Defense Weapon was arrived at hastily should consider the following:

The U.S. Government first tested a 9mm semiauto (Luger) in 1903 and a double-action semiauto (Knoble) in 1908. It has been "testing" ever since.

It all started in 1901 when 1,000 .30 caliber (7.65mm) Lugers were purchased and allocated to troops for testing. Reports from troop commanders trickled in but weren't good. They felt the small projectiles lacked "stopping power," as did bullets of the .30 caliber Mannlicher and Mauser pistols which also had been briefly tested.

The bias in favor of larger pistol projectiles came from the ongoing Philippine insurrection where the recently adopted and underpowered .38 Colt double-action revolver (in .38 Long Colt) failed to "give a good account of itself" against the Moros, thus prompting a recall to duty of the .45 Single Action Army revolver. It should be noted, however, that the .30-40 Krag rifle didn't always "stop" a drugged and/or fanatic Moro, either.

In April 1903, the Army proposed trading 50 of its test Lugers to Deutsche Waffen und Muni-

tionsfabriken (DWM) for 50 "bigger-is-better" 9mms. Georg Luger himself brought sample 9mms of several barrel lengths to the U.S. the following month, and the testing proceeded.

In 1904, the Army established a board consisting of Col. Louis LaGarde, Medical Corps, and Col. John Thompson, Ordnance Department, to look into the caliber matter. As reported in Col. LaGarde's "Gunshot Injuries" of 1914, the board shot a number of different cartridges, including the two Lugers, several .38s, several .45s, and even some .476 caliber. The board shot into 10 cadavers, 16 beeves, and two horses and decided that the best cartridge for "stopping power" was the big .476 caliber at a moderate 729 fps, with smaller sizes less effective, and recommended the .45 caliber as the minimum. *But* all the Luger bullets tested were jacketed, and all the revolver bullets were of unjacketed soft lead that were soon to be considered unacceptable for military use. Despite this anomaly, the Army went back

This article first appeared in American Rifleman. *Cartoons reprinted with permission from* Bill Mauldin's Army, *by Bill Mauldin, 1983, Presidio Press, 31 Pamaron Way, Novato, CA 94947.*

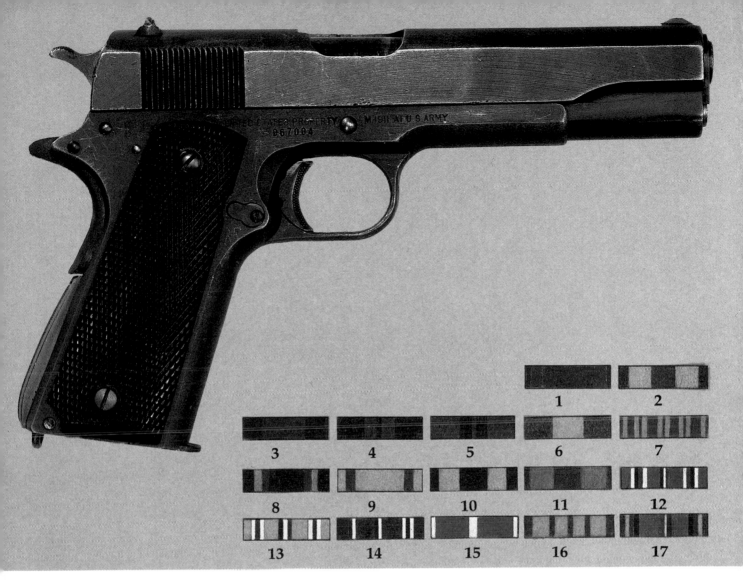

These are campaign ribbons bestowed on American veterans of four major 20th century conflicts, in which the M1911A1 also served. The ribbons are from World War I, World War II, Korea, and Vietnam—and include the following: 1. Philippine Campaign, 2. Mexican Campaign, 3. Nicaragua 1912, 4. Haiti 1915, 5. Dominican Republic 1916, 6. Mexican Border, 7. Nicaraguan Campaign 1927–1933, 8. Yangtze Service, 9. China Service, 10. U.S. Navy Expeditionary, 11. USMC Expeditionary, 12. WW II American Campaign, 13. WW II Asia-Pacific, 14. WW II Europe-Africa—Mideast, 15. Korean Service, 16. Republic of Vietnam Service, 17. Armed Forces Expeditionary.

to the .45 Colt cartridge in the Colt Double Action Frontier model, a rod-ejector, for use by the Philippine constabulary, and, in the Model 1909, issued as a military version of the Colt New Service .45.

The Colt .45 revolvers were, however, interim solutions only. The Army really wanted a semiautomatic. Except for its .38 caliber, the Army liked the Model 1900 Colt semiauto used in the 1904 tests.

In 1907, the Army requested more pistols and revolvers for testing, but all were intended to use the same government-supplied ammunition, differing only in their rimmed or rimless treatments, with 230-grain .45 caliber full-jacketed bullets traveling at about 800 fps. The speedy 9mm was temporarily forgotten in the post-Moro environment.

Colt and S&W provided revolvers, DWM sent in a .45 Luger, and other .45 semiautos arrived from Savage, White-Merrill, and Bergmann. There also came two versions from Knoble (one was a *double action*, a Webley-Fosbery auto revolver, and, last but not least, the Model 1905 Colt.

The Model 1905 .45 Military Automatic Pistol used in the tests was similar in appearance to and generally based on John Browning's Model 1900 .38 automatic. It was soon to be tested in the improved Models 1907, 1909, and 1910.

It was apparent that the Colt had the edge over its closest competitor, the Savage.

Final tests were conducted in March 1911, between the Savage and the Colt. The Savage had some malfunctions and broke a few parts, but the Colt went through the 6,000-round test without a malfunction.

This gun, chambered for Browning's original caliber .45 Colt Auto cartridge, was adopted as the standard U.S. service pistol, the Model 1911. Around 1924, the gun was modified by the addition of a longer spur on the grip safety, an arched mainspring housing, and by cutting away a portion of the receiver just back of the trigger. This, then, became the U.S. Pistol, Automatic, Caliber .45 M1911A1, which has served faithfully up to the present time with essentially no change. In commercial form, the gun has also been chambered for the .38 Super cartridge, the 9mm Luger (and other metric calibers for export), and the .22 Long Rifle (plus conversion kits). This gun has been extensively used by civilian, police and military, shortened, lengthened, lightened, heavied, accurized, otherwise gunsmithed and extensively copied, to the eternal credit of John Browning.

The new handgun was put to the big test when the U.S. entered World War I in 1917. The story is best told by the enthusiastic Benedict Crowell, Assistant Secretary of War, in his "America's Munitions 1917-1918": "The American pistol was one of the great successes of the war. For several years before the war came, the Ordnance Department had been collaborating with private manufacturers to develop the automatic pistol; but none of our officers realized, until the supreme test came, what an effective weapon the pistol would be in the hand-to-hand fighting of the trenches. . . . In the hands of a determined American soldier, the pistol proved to be a weapon of great execution, and it was properly feared by the German troops.

"Only a few men of each infantry regiment carried pistols when our troops first went into the trenches. But in almost the first skirmish, this weapon proved its superior usefulness in trench fighting," Crowell continued.

"By midsummer of 1917, the decision had been made to supply to the infantry a much more extensive equipment of automatic pistols than had previously been prescribed by the regulations— to build them by the hundreds of thousands where we had been turning them out by thousands."

In December 1917, orders were given Remington U.M.C. to make 150,000 pistols. But it was difficult to get proper drawings from Colt, as the gun had long been made by old hands in the plant who worked up little tricks to make the job easier or the parts fit better. In short, the pistol did not really match the drawings.

But, finally, revised drawings were made and in the summer of 1918 contracts for the M1911 pistol were also let to National Cash Register Co., North American Arms Co. of Quebec, Savage Arms Co., Caron Bros. of Montreal, Burroughs Adding Machine Co., Winchester Repeating Arms Co., Lanston Monotype Co., and Savage Munitions Co. None of these firms had ever made the pistol before, but all were working hard on getting into production when the Armistice stopped it all. The only guns actually used were made by Colt, Remington U.M.C., and Springfield Armory.

According to Crowell, when the war began, the Army alone owned approximately 75,000 M1911s. At the signing of the Armistice, there had been produced and accepted since April 6, 1917, a total of 375,404 pistols. In the four months before November 1918, the daily production averaged 1,993 pieces at an approximate cost of $15 each.

In World War II things were much better organized, and production on most small arms got underway fairly quickly in 1940-42, because the basic designs and manufacturing techniques had been worked out and standardized long before the outbreak of hostilities.

Pistols were made at first only by Colt, but in 1942 three other producers were added—Remington Rand, Union Switch & Signal, and the Ithaca Gun Co. Even the Singer Sewing Machine Co. made a few. Pistols were in short supply during the whole war, their production suffering from the serious obstacles of low priorities and fluctuating requirements.

While adoption of the M1 carbine brought some reduction in overall pistol requirements, the demand nevertheless remained strong throughout 1944. Almost two million pistols were produced during the war, plus almost a million .38-caliber revolvers.

At that time, the Ordnance Corps was responsible for developing (and manufacturing) weapons for the units using arms—Infantry, Cavalry, Air Corps, Artillery. Sometimes the item was developed to meet Military Characteristics (MCs) sent in by the customer. These told the Ordnance folks what the customer felt he needed in the way of weight, rate of fire, dimensions, ease of use, capability, and so on. The MCs constituted a "wish list" and were usually tough to meet, as the customer was pushing beyond the "state of the art." On other occasions, Ordnance would propose something which had come from ideas in-house or from outside suggestions, to see if the user wanted it. All this was a considerable improvement over the system in effect at the turn of the century, where the Ordnance Department developed things and then made the decisions on their adoption for use.

Toward the end of World War II, many extended meetings were held to get the opinion of the users on how their equipment had worked, what improvements could be made and what new equipment was needed. It was interesting to hear the many violent differences of opinion, depending on whether the man had been in the jungles of the Pacific, the mountains of Italy, or the plains of France.

The Infantry Board studied future weapons for the Infantry, and in 1945 wrote MCs for a new pistol. As a Test Officer and later Test Director of the Board, I recall arguing over many of the features with the project officer. It was to be 9mm in caliber, light in weight, and have the double-action feature. Other details I have forgotten, but the MCs were recorded by the Ordnance Corps, to enable them to start development work on guns to meet these requirements. A year or two later, I transferred from the Infantry to the Ordnance Corps and went to the Small Arms Research and Development Branch in Washington. Lo and behold! One of the jobs that fell into my lap was the new lightweight pistol. As I recall it, High Standard had a contract to work on a model. We had found a bug in the gun, and on a visit to the plant I told them that their gun had a safety problem. After some heated discussion, one of the engineers quietly left with the gun. Pretty soon we heard the sound of a shot and he shortly came back in with a sheepish look on his face.

Meanwhile, Smith & Wesson was developing a gun in-house to meet the same MCs. The first gun they sent down for us to work with looked fine, but it also had a safety problem. They had a magazine disconnect which pulled the ejector out of place. So if you had a loaded chamber and took the magazine out, you could work the slide all you wanted, but you would never eject the loaded cartridge. It would ride back and forth with the slide, putting the cartridge back in the chamber each time. When I called them, they were much surprised and didn't know how they could have missed this in all their testing. But

that gun grew into the fine Model 39 and its later versions.

But the enthusiasm for developing a new lightweight pistol disappeared when the Army looked at the stockpile of .45s and what it would cost to replace them, their spare parts, ammunition, holsters, and other equipment, compared to the benefits to be gained. As a result, Ordnance Corps work was stopped on this project in the mid-50s.

For nearly 70 years, Aberdeen Proving Ground has been the center of Army engineering and development testing of weapons of various sorts. Many ranges, elaborate instrumentation, and experienced engineers and testers made this an outstanding installation. Guns were tested there from time to time during their development at or through the arsenals and armories with detailed and accurate instrumentation. When the gun had seemed to be satisfactory, it was normally sent for user testing to the customer for its evaluation. In the case of small arms, this was usually the Infantry Board at Ft. Benning, where tests were conducted using soldiers, to see if the item could be handled properly by the user and to see if it really had any value to him. I spent four years there doing this delightful and fascinating testing.

Aberdeen was an interesting place, and for a gun enthusiast to be assigned there was like turning a 12-year-old loose in a candy store. I know. I was assigned there for three years after I transferred to the Ordnance Corps from the Infantry.

While at Aberdeen, I ran into another ardent shooter, one Bill Brophy, a fine shot and eager experimenter. For several years, I was also assigned as the Ordnance Officer of the National Matches at Camp Perry, and for a good part of the time Bill worked with me, taking over the job when I got kicked upstairs to be the Match Director for four years.

"Didn't we meet at Cassino?"

"Pass, friend."

"What say we go down an' strafe 'em?"

After the Korean War hiatus, the matches resumed in 1953. We issued M1 rifles and M1911 pistols to the competitors, to a certain amount of unhappiness. The shooters felt that the M1 was "not up to the good ol' Springfield 03" as a match rifle, and "knew" that the issue .45 wasn't accurate. We got Ordnance interested in showing that the M1 and the .45 could be made into fine match guns with a little work. In the process, Bill and I wrote an article for the *American Rifleman* (August 1959), giving results on the testing (mostly by Brophy) of .45 ammunition and of hundreds of .45 pistols, straight issue as well as guns worked over by gunsmiths.

We, and gunsmiths in the field, found that the very things which made the .45 a reliable combat gun—the loose fit of the parts—ruined accuracy. Most of the "accuracy jobs" tightened the slide on the receiver, the barrel bushing in the slide, the barrel locking in the slide and some other loose fits, all trying to make the critical parts return to the same place each shot. If the work was done properly, the result was a fine-shooting gun.

I could feel for Frank Allen, a pistol-shooting Air Force colonel when he was assigned to Aberdeen. His enthusiasm showed itself in a March 1952 *American Rifleman* article covering tests run on various auto pistols. Included were the M1911A1 and a number of foreign pistols in various calibers.

All were run through an endurance test of 5,000 rounds—or as far as they would go. Most of the guns gave many malfunctions or finally quit altogether. In contrast, the .45 sailed gaily through the 5,000 rounds with only one minor stoppage.

Some 15 years ago, the U.S. Treasury Department had tests conducted by a private testing laboratory on a number of handguns in a fruitless search for the elusive "Saturday Night Special."

The tests ran over an eight-month period and covered some 150 guns of 38 models, from .22 Short revolvers to the .45 auto. Of interest here are the results with the bigger pistols—9mm and .45, but none of them were double actions. In the endurance test, the lab tried to put 5,000 shots through each of two samples, with one gun getting an initial proof round and the other a proof round after every 100 shots, something you would never expect to encounter in actual use.

Incidentally, the Sporting Arms and Ammunition Manufacturer's Institute's standards call for maximum average pressure of 19,900 c.u.p. in the .45 and a nominal 38,500 c.u.p. for the proof load. Similar figures for the 9mm show 35,700 c.u.p. for service loads and 47,500 for proof.

Several 9mms and .45s of U.S. and foreign make were run through the tests. The only ones which went all the way through the 5,000 rounds were the two Colt .45s, one Browning Hi Power 9mm,

Simple disassembly, sturdiness, and a minimum of parts have always distinguished the Colt Government .45 and its civilian versions. Both of these pistols are Colt MK IV Series '70—one gun assembled and one disassembled. (Randy Lawson/National Rifle Association photo)

and one of the 9mm French MABs. Another Browning almost went 5,000 (with 47 proof rounds) and a second MAB went over 4,900 with 50 proof rounds. This test, plus the 1911 acceptance tests and the Aberdeen affair, all showed that the M1911 was a mighty reliable pistol.

The last M1911 pistols were bought in 1945, but even before that, many M1917 Colt and S&W .45 revolvers had been issued, together with .38 Special revolvers for special purposes. In 1963, the Air Force switched from the M1911 to the .38 Special revolver almost entirely, and since then the services have bought many kinds of revolvers, and ammunition to match. By 1976, testing started out again in confusing earnest. Something had to be done, for the old M1911s were wearing out and the .38s were of too many types for logistical practicality.

In that year, the Air Force Armament Development Laboratory at Eglin A.F.B., Florida, began to evaluate commercially available 9mms with a view toward replacing *all* the .38s in USAF inventory. The 9mm pistols tested included double actions by FN, H&K, Star, S&W, Colt, and Beretta, with the last rumored to have taken the lead.

In 1977, the Department of Defense asked the House Appropriations Committee for some money to design a new .38 cartridge. Instead, the committee asked a subcommittee to investigate the existing status of military handguns.

In 1978, that subcommittee reported that there were something like 25 different handgun models and 100 different ammunition items in military stocks. This seemed a trifle excessive, and the military was directed to standardize.

The Air Force tests were incomplete and inconclusive, and in late 1978 the Department of Defense placed the responsibility for small arms research development, testing and evaluation under the authority of the Joint Services Small Arms Program.

A short time later, the Under Secretary of Defense directed the establishment of a joint study to determine the minimum number of ammunition types, what to do about the handgun problem, and if all the U.S. services should adopt the 9mm or stay with .38s and .45s. Everyone joined in this JSSAP study—Army, Marines, Navy, Air Force, Coast Guard, FBI, Secret Service, and others, as everyone had an interest and all had experience to contribute. The Army chaired the study.

A report of 1980 showed that 70 out of 95 ammunition items had been eliminated. A number of the 25 remaining were .38 caliber, mainly used for Air Force revolvers. If one caliber was adopted for all services, it was obvious that the number of ammunition items could be still further reduced.

After much debate, the JSSAP 1980 study recommended a 9mm Luger/Parabellum/NATO handgun family, with a basic gun for general use and limited issue of a smaller concealable version. This would pretty well take care of everyone's requirements and called for more tests.

In 1981, the Army issued a request for proposal that it hoped would result in a contract award for the so-called XM9 pistol in January 1982. But that same year a congressional investigations subcommittee of the Armed Services Committee held hearings to examine the requirements for the 9mm gun, why it was preferred to the guns in the inventory, the cost of the changeover, the specifications of the new gun, and the ability of U.S. firms to make it.

The General Services Administration also had decided to have another look at the new proposal. Was it really necessary to change calibers? If so,

couldn't existing guns be converted to 9mm at less cost? In fact, GSA had a M1911A1 converted to 9mm and found that it worked OK. This shouldn't have come as any great surprise, since Colt had made the 9mm on the M1911A1 action for some time. Although the GSA proposal to convert .45s to 9mm would be cheaper, it didn't make too much sense, since there were not enough .45s left to convert for all hands, and those remaining were old and tired. It would cost about $160 to refurbish an old pistol into a relatively unsatisfactory, partly new pistol; would not give the troops what they wanted in the way of magazine capacity, double action, safety, light weight, and other features; would not standardize on a single gun for the services, and would leave the question of the .38 revolvers up in the air.

In 1981, pistol makers were invited to submit 30 pistols for evaluation in the fall of that year, with production of the selected model scheduled for 1983. Models were submitted by several makers, but early the next year, after some testing, the whole thing was called off, as none of the guns gave desired reliability. More testing loomed.

After this affair, the House Committee on Appropriations had some thoughts on the subject of all that testing with no results and made dire threats. Not too strangely, this stirred up some prompt action. Some revisions were made to the Joint Services Operational Requirements (what we used to call "Military Characteristics") and *new* trials were started, with eight makers each submitting 40 guns designated "XM9": Colt, Walther, S&W, Steyr, Browning, SIG, Beretta, and H&K.

Two guns successfully passed these tests, SIG and Beretta. Based on the "lowest overall costs," Beretta was announced the winner on January 14, 1985, and a contract was subsequently signed.

Beretta began by importing made-in-Italy pistols, and then began a transition to American manufacture by importing parts from Italy for assembly here. By the third year, they expect to be in full operation in Maryland.

There will be those who view with high joy the demise of the .45 as our military handgun; folks who have cursed the weight of the gun, the large size, the heavy recoil and the difficulty of shooting it. They will laugh with glee as they throw their detested .45s in the junk box and grab the new guns.

And there will be those who have seen the M1911 develop into a top-grade match pistol, with excellent functioning, who have felt the weight of the gun on their hip as a very comforting thing when trouble brews, have used the gun in combat and have faith in its reliability, and who have great confidence in the .45s man-stopping power. They will join a large group which reluctantly says, "Goodbye, Old Friend!"

Hand Cannons for Big and Dangerous Game

J.D. Jones

My perspiration-dampened shirt suddenly turned my back chilly, telling me I was in trouble. Because of a wind shift, we were now less than 40 yards upwind from the elephant herd. The lead cow caught our scent a few seconds later and charged without hesitation or a sound. She was amazingly fast and came at me from the left. I felt the fear and considered the possibility of running, but my preconditioning paid off. I immediately put aside those impractical thoughts and made up my mind to shoot if she came past a certain tree about 20 yards in front of me. The elephant approached the tree directly downwind and turned dead on while reaching for me with her trunk. I had the .375 JDJ single-shot handgun up. I put the crosshairs of the 2× Leupold between the elephant's eyes and a foot below them on the trunk, and I touched off a shot.

The elephant stopped instantly and collapsed to her left. At the shot, I broke open the rebarreled Thompson/Center pistol and had another round in it before she hit the ground. There was no need to fire again, though. She was dead. The 300-grain Hornady solid bullet had entered her trunk, penetrated upward through the top of a tusk, through the skull and lower right portion of the brain. It exited the skull, and we later found it in the neck muscles, about a foot behind the skull.

That self-defense elephant kill and many other kills demonstrate that some handguns are enough gun for hunting truly big game. Let's look at some of the more suitable guns and ammunition for hunting medium to big game.

Most handgun hunters are deer hunters, and quite a few of us hunt antelope and feral hogs. A surprising number hunt black bears. A small percentage of us hunt elk, moose, large bears,

caribou, and goats. I strongly suspect that more of us hunt in Africa than in Alaska. Whatever and wherever, the trend toward hunting big game with a handgun is growing rapidly. Let's explore in some detail the calibers and guns that are suitable for serious hunting.

REVOLVERS

In the opinion of many handgun hunters, with the right ammunition and shot placement, revolvers are capable of killing any animal. These handguns are fast, comparatively light, and easy to carry. They are very much at home in woods and thick brush where shots are close and fast. The best revolvers are limited to an approximate accuracy level of five-inch, 100-yard five-shot groups. Scope sights improve your ability to hit what you shoot at, just as they do with rifles. Almost all of them, however, need work by a gunsmith to improve the trigger pull before they are capable of doing a good job in the field. Colt and Charter Arms are limited to .357 Magnum caliber as effective deer cartridges. Dan Wesson, Smith & Wesson, Ruger, Interarms, and several other revolver manufacturers chamber the .357, .41, and .44 magnums. Freedom Arms chambers the .454 Casull, which is the most powerful factory revolver cartridge now made. In my opinion,

Top to bottom: Remington XP-100 bolt gun, which is often rebarreled to fire heavy hunting loads; Thompson/Center Contender; as rebarreled and customized by the author; Wichita Arms break-action single-shot Hunter International; Smith & Wesson Model 29 revolver in .44 Magnum; Ruger Redhawk, also in .44 Magnum. (Stanley W. Trzoniec photo)

This article first appeared in Hunting Guns

the Freedom Arms revolver is the highest-quality revolver currently available.

No cartridge developing less power than the .357 Magnum should ever be considered a deer cartridge. Just because grandpa once killed a deer with a .32/20 lead-bullet factory load from a revolver doesn't make it an effective sporting cartridge for deer.

The .357 Magnum: The .357 is the bottom rung of the ladder of deer cartridges. Few people with much hunting experience continue to use it. The .357 is capable of killing any deer that ever walked—but only at close range. I've used and seen the .357 used enough to know that shot placement must be exact or the shooter is in for a heap of trouble.

Elgin Gates once took a Cape buffalo with a four-inch .357 revolver, using Winchester-Western metal-piercing ammunition. Elgin didn't like the thorn tree that the buffalo put him into, so he settled the matter with a single brain shot. But the range was only a few feet.

Generally speaking, 110- and 125-grain .357 bullets are poor choices, with the possible exception of using them on small deer or antelope. Even then, if you have to shoot at an animal going away, you are in trouble because of the limited penetration. Light-bullet ammo in this caliber is usually intended for self-defense against criminals. I've seen some of the 125-grain hollow points disintegrate and fail to penetrate more than three inches in wild hogs. I much prefer bullets in the 158-grain or heavier class in either jacketed or cast versions. I believe in getting as large a hole all the way through the animal as I can. It lets more blood out and air in, resulting in easier-to-follow blood trails and quicker kills.

In handguns, heavy bullets penetrate better than light ones. Even if a broadside shot is presented and a perfect lung shot made, the animal usually runs. If the animal doesn't drop instantly, I shoot again if I have a fair chance of hitting. The shot that you think was a perfect lung shot frequently turns out to be a liver shot and, although quickly fatal, a deer can go a lot farther with a liver shot than with a lung shot.

The .357 Maximum: Currently, the only revolver chambered for this Remington round is the Dan Wesson. This cartridge gives performance in a revolver approximating that of the .357 Magnum when fired in the 10-inch Thompson/Center single shot. In my opinion, it's little better than the .357 Magnum.

The .41 Magnum: It's a lot more gun than the .357 and makes a lot more sense than the .357 for deer hunting. Its recoil approximates that of the .44 Magnum more than the .357, and most people choose the .44 for the sake of greater power. I've never been able to get the velocities from the .41 that the .44 produces with equivalent bullet weights. That doesn't make any difference, though. The .41 caliber bullets seem to shoot flatter at long range than equivalent weights in the .44. The Sierra 170-grain .41 is the lightweight speed demon in this caliber and does well on small deer. Again, I prefer the 200- to 210-grain bullets in this caliber. The 210-grain factory loads are fairly good game stoppers, although handloading can improve the .41's ability in the field by a considerable amount.

The .44 Magnum: This is undoubtedly the undisputed king of the revolver cartridges for hunting. The .44 is capable of handling readily available bullets ranging in weight from 180 through 320 grains. Its case capacity is reasonably large and bullet selection is truly great for handloaders. It does outshine the .41 in versatility and absolute ability in the field. This, however, is an unimportant point until we get well above deer in animal size. On most game that I've taken and seen taken with the .44 Magnum, I've always had the feeling that the .41 would have done practically the same job with the same hit. Only on large game have I felt that the advantage of a .44 was really significant. The advantages of the .44 are real, however, and I can't think of any valid reason to buy a .41 instead of a .44. The .44 will do anything that the .41 will do and will surpass it by a considerable margin in an all-out power race—at least on paper. For truly large game, such as moose, eland, Cape buffalo, and even elephants, the .44 has been the only revolver cartridge to warrant consideration. It is at its best for really large animals with handloaded 320-grain JDJ cast bullets, which will exceed 1,400 fps from any decent .44 with a barrel length of 7.5 inches or more, and will provide greater penetration than any other bullet in .44 caliber fired from a revolver. Last summer, Larry Kelly, head of the Mag-Na-Port company, took a trophy bull elephant with a heart shot with this bullet in a Stalker Ruger custom-conversion revolver. The distance was close, as is usual in hunting dangerous game, and a single heart shot dropped the gigantic beast within 40 yards. Others have taken numerous Cape buffalo with this bullet in the .44 Magnum revolver.

The .454 Casull: Available in the excellent Freedom Arms revolver, this factory cartridge generates significantly more power than the .44 Magnum. In my opinion, it's a waste to use it with any bullet weight under about 270 grains, and it's much better with 300- to 340-grain handloaded bullets. There is no doubt about it, this one is a handful with heavy hunting loads that may range from 62,000 to 65,000 copper units of pressure (c.u.p.), which makes it the current champ among revolvers as far as handling high pressure and being the most powerful production revolver cartridge. With this cartridge and gun, we step onto the threshold of shooter tolerance. Being rather light in weight for such high power, it produces a big recoil with maximum loads. However, I feel that almost anyone can learn to shoot a .44 Magnum with full-charge loads, assuming that the individual is physically large enough to hold the gun. (Rosetta Sliva, at a whopping 86 pounds, is a consistent silhouette winner with a .44 Magnum Super Blackhawk.) Many individuals who are comfortable with the .44 will not, however, be able to control and shoot effectively with the .454 and maximum loads. This gun is so new that little meaningful testing has been done with it outside the factory. I designed a 340-grain cast bullet for it but have not done any significant testing as yet.

BALLISTICS OF SOME HANDGUN CARTRIDGES

Caliber	Bullet Weight (grains)	Nominal Velocity (average muzzle velocity in fps)	Energy (muzzle energy in foot-pounds)
6.5 mm JDJ	120	2,400	1,533
6.5mm JDJ	129	2,300	1,514
.309 JDJ	165	2,350	2,022
.357 Magnum	158	1,300	592
.41 Magnum	210	1,300	787
.44 Magnum	180	1,600	1,036
.44 Magnum	240	1,350	971
.45 ACP	185	940	363
.454 Casull	260	2,000	2,308
.454 Casull	340	1,550	1,812
.30 Herrett	125	2,250	1,405
.30/30 Winchester	150	2,150	1,539
.357 Herrett	158	2,400	2,000
.35 Remington	180	2,300	2,113
.358 JDJ	250	1,970	2,152
.375 JDJ	220	2,200	2,362
.375 JDJ	270	1,970	2,324
.375 JDJ	300	1,940	2,511
.411 JDJ	300	1,950	2,532
.411 JDJ	400	1,700	2,564
.444 Marlin	240	1,900	1,922
.45/70 Gov't	300	1,650	1,812
.45/70 Gov't	400	1,650	2,416
.45/70 Gov't	500	1,425	2,250

OTHER REVOLVERS

It's true that the .44 Special, .45 Colt, and a few other revolver cartridges can be handloaded to effective velocity and energy levels for deer hunting. These souped-up revolvers are reasonable as 100-yard deer guns, and experienced long-range shooters will stretch that distance somewhat. Silhouette shooters regularly take 10 out of 10 rams at 200 meters with such revolvers. This feat, however, is accomplished on a range and at known distances, not in the hunting field. Any of these handloaded calibers will harvest game. Just remember—the bigger the hole through the animal, the more effective the hit.

AUTO PISTOLS

At this point, the LAR Grizzly in .45 Winchester Magnum is the only commonly available auto pistol suitable for big-game hunting. It lags behind the .44 Magnum in just about everything, but it will take medium game effectively at short ranges. In my opinion, the .45 ACP isn't enough gun. I've used it and feel that its maximum range for deer-class game is 50 yards with heavy handloads. Shot placement, therefore, is very critical. Several years ago, I shot a Corsican sheep with a heavy handload in the .45 auto, and it took off like a scalded dog. I followed up closely but I couldn't see any sign of a hit, so I assumed that I had missed. I heard a shot a few minutes later and went to check it out. Another hunter had "my" sheep. All I had done was bounce a bullet off the base of its horn. Apparently either the ram moved its head just as I shot or I messed up myself. The bullet left a lead streak about two inches long on the horn. The wildcat .41 Avenger conversion for the 1911 Colt is a better hunting rig than the .45 ACP, but that isn't saying much for it.

Contrary to what you read from time to time, I've never really seen much evidence of "knockdown" or "shock" power on big game caused by any revolver hit. Obviously a brain or spine hit will drop an animal instantly. A bullet, traveling at revolver velocities, depends on penetration of vital organs to kill. The animal dies either because of interruption of the nervous system through a brain or spine hit, or because of the lack of oxygen caused by a lung, heart, or artery hit. In most cases, an animal hit in the heart or lungs will run until it can't go any farther. I have never seen an animal knocked down by a chest hit made with a revolver. Dropped instantly by a spine hit—yes. Dropped by a broken shoulder hit—yes. Dropped by a lung or heart shot—no. This is one of the reasons why a good blood trail is important. In many areas, a deer that travels 50 yards and holes up before dying is lost. Many lung- or heart-shot deer that are hit with high-powered rifles run a ways, too.

BOLT-ACTIONS

The Remington XP-100 bolt-action single-shot handgun was the first successful modern single-shot pistol. It's still made by Remington in the original .221 Fireball varmint caliber. Remington also chambers it to fire the disappointing 7mm BR wildcat, which is essentially a silhouette-competition cartridge. The ammunition is difficult to make, dies are expensive, and it therefore makes a lot more sense to convert the XP-100 to 7mm-08 or .284 Winchester. Custom conversions of this handgun can handle almost any cartridge. I find that .308 Winchester is the most popular conversion, closely followed by several wildcats in .375 and .45 caliber. The trend is definitely to bigger bores for hunting. More power can be packed into an XP-100 package than any other type of handgun action, but it's fairly expensive to do. Single-shot bolt-action handguns in many factory and wildcat chamberings and varying configurations are also made by Wichita Arms in Wichita, Kansas.

THOMPSON/CENTER CONTENDER

Chambering the .30-30 in the 10-inch Thompson/Center gave rise to the .30 and .357 Herrett wildcat cartridges, based on the shortened .30-30 case fireformed to a more modernistic shape. The Herretts are still very effective game getters in the break-action Thompson/Center. In a 10-inch barrel length, the .44 Magnum and the Herretts are all that are needed to handle just about any deer or hog hunting. The .357 Herrett and .44 Magnum will do well if loaded with the right bullets on elk- and moose-sized animals, assuming that the range is kept relatively short, only good shots are taken, and the shooting is accurate. Although the Thompson/Center is available in more than 20 calibers, the above three cartridges will ballistically do just about everything for deer hunters. Handloaded to Herrett pressures, the .30-30 will outperform the Herrett. The exceptional accuracy of these cartridges in the Thompson/Center make 200-yard shots on deer and antelope feasible for really good shooters. Many scoped Thompson/Centers will consistently shoot 1.5-inch 100-yard groups.

If you look through the latest edition of *Gun Digest* and the *Shooter's Bible*, you will turn up other single-shot handguns and revolvers that are useful for hunting as well as for silhouette shoot-

ing. It pays to do so in order to be aware of everything that is available. If you find something that you like, make further inquiries about it and later, perhaps, order the gun through your dealer. Never restrict yourself to the limited variety of handguns to be found in the average gunshop.

After having used many different cartridges and guns for years, I became well aware of their shortcomings and their strong points. Successes were sweet and, for normal woods hunting for deer and hogs, I was satisfied for years. I also passed up a lot of shots that I knew were marginal with the calibers that I was shooting and went home empty-handed. Hunting time became scarcer and costs rose, too, so I decided to do something about it and developed the JDJ series of cartridges and cast bullets. My shop now adapts Thompson/Center Contender and XP-100 handguns to fire these as well as other cartridges—wildcats and factory loads.

My heavyweight cast bullets for revolvers provide excellent penetration. The 320-grain .44 Magnum possesses enough penetration to take heavy game. I certainly don't recommend the .44 for the really big stuff, but it has taken just about everything. My T'SOB scope mount is a really reliable scope-mounting system for handguns. In my experience, the JDJ series of cartridges provides exceptional accuracy and killing power at both short and long ranges. The .375 JDJ cartridge for single-shot pistols makes hunting the heaviest game in the world with a handgun a practical undertaking. SSK barrels, which I furnish, in calibers such as .300 Savage, .30-40 Krag, 8×57 Mauser, .444 Marlin, and .45-70 provide good hunting rounds for hunters who do not reload and who must depend on factory ammunition, but performance of these rounds can be improved by handloading.

The smallbore JDJ series is based on strong .225 Winchester brass maximized in case capacity. Forming the case is accomplished simply by expanding the neck to the appropriate caliber and fireforming the case. No trimming, reaming, or other work is necessary. It is available in the .226, 6mm, .257, 6.5mm, .270, and 7mm JDJ designs. All of them have their particular strong points but, as an all-around hunting cartridge, the 6.5mm JDJ is the best of the lot. It will push the 12-grain Speer Hot-Cor bullet to more than 2,400 fps and will group around an inch at 100 yards. Its velocity retention is excellent. The bullet expands without weight loss and penetrates exceptionally well. The 125-grain Nosler Partition and 129-grain Hornady bullets also perform well. There is nothing magic about the 6.5mm. The answer to its effectiveness is simply the fact that the bullets are being driven at velocities suitable for their construction. The first 100 deer and antelope to fall

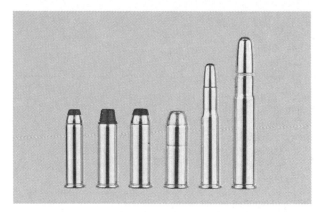

Some hunting handgun loads shown here from left include .357 Magnum, .41 Magnum, .44 Magnum, .454 Casull, .30–30 Winchester, and .375 JDJ, based on the .444 Marlin necked down to .375 caliber. (Stanley W. Trzoniec photo)

to the 6.5mm JDJ were completely penetrated. Not one bullet was recovered.

The 6.5mm JDJ gets into the realm of knockdown or shock power due to the higher velocities, combined with reliably expanding bullets. We have seen antelope and deer go down instantly, on occasion, with lung shots. Mark Hampton and I used it extensively in South Africa on plains game weighing from about 80 pounds (springbok) to around 400 pounds (Cape hartebeest) at ranges of 220 to 304 paces. Twelve animals were dispatched with a total of 14 shots. The 125-grain Nosler bullet was used exclusively on this hunt, and its performance was excellent. Three bullets were recovered, and all showed typical Nosler performance. Sighted-in three inches high at 100 yards, the 6.5mm is dead on at about 260 yards. This flat trajectory allows for exceptional accuracy of shot placement from converted handguns. Certainly, we were careful of our shooting and took a solid rest for every shot. In open country, it simply isn't possible, in many cases, to approach close enough for a revolver shot. Had we been using revolvers, I expect that we would have probably taken about four of the animals in the same time period.

The larger series of JDJ cartridges is based on the .444 Marlin case necked to a different diameter. The .375 JDJ is a versatile handgun hunting cartridge. It's formed by simply necking the .444 case to .375 in the full-length sizing die and loading with a full charge of powder. You are then ready to go hunting. The .375 JDJ provides good accuracy with the power needed to cleanly take the largest animals. It will push a 220-grain Hornady at more than 2,200 fps, and 270- and 300-grain bullets in excess of 1,900 fps. Sighted-in with the 270 Hornady three inches high at 100

yards, it is about five inches low at 200. At 200 yards, it consistently penetrates elk completely if broadside hits are made. In 1984, 12 elephants were taken with this cartridge. In addition, it took numerous Cape buffalo, leopards, lions, and a few rhinos. In the United States, it has taken large numbers of deer and antelope, numerous elk, several sheep, several moose, and two brown bears. The .375 JDJ's effectiveness is due, in large part, to the excellent factory .375 bullets available to handloaders.

Adapting the .45-70 to the Thompson/Center Contender was a natural progression of ideas. The .45-70 can be made to sit up and talk in the SSK-barreled Thompson/Center. A 400-grain Speer will do 1,650 fps from a 12.5-inch barrel and a 500 Hornady will hit more than 1,400 fps. Ray Guarisco took an elephant with one last summer. I've gotten up to seven feet of penetration in large African animals with it. With the 400-grain Speer bullet, it is devastating out to 150 yards. The Speer bullet is soft and expands readily. It is a sure deer dropper. I don't know of any deer hit with it that went more than 10 yards. The 500-grain bullets do not expand on game, but they will shoot lengthwise through a concrete block or completely through a hippo's head. A few years ago, I was in a party tracking a wounded Cape buffalo. The .45-70 was loaded with Hornady 500-grain solids. Visibility was very poor, and I noticed that I couldn't see the buffalo's horn when I spotted him in the bush, angling toward me at about 25 yards. At that range, I wanted to make a spine shot. I fired, fully recognizing the possibility of hitting the horn. The buffalo rocked backward at the shot and a spot of white showed that I had hit horn. The next shot went over the horn and into the spine, taking him down instantly. Most impressive was the first shot. It penetrated the horn, hit the shoulder sideways, took out a couple of ribs, went through the top of the lung and the liver, and was found just inside the stomach. All of the high-performance cartridges for Thompson/Center and XP conversions are available from SSK Industries, Rte. 1, Della Dr., Bloomingdale, OH 43910 (614–264–0176). The high-power calibers are called Hand Cannons. Complete guns or custom barrels for Thompson/Center are available, as well as T'SOB scope mounts for handguns. I head this outfit, as well as being the director or Handgun Hunters International, an organization dedicated to furthering the sport of handgun hunting.

Recoil is always something that comes to mind when large-caliber guns are discussed. Sure they kick. Handguns, like rifles, must be set up right or they will really beat you. To handle the recoil, don't go rigid just before you fire, as you would do with a rifle when firing at stationary game.

Mauser "Broomhandle": from the Boer War to Today

Garry James

There are some arms, despite antiquated systems or calibers, that seem to defy time. In a few very special cases, form outstrips function and a "classic" is born. The C/96 "Broomhandle" Mauser is an archetypical example.

Despite a design rooted firmly in the late 19th century, this 10-shot auto pistol remains for some (including myself) one of the most aesthetically pleasing handguns of all time. Although the gun was a product of Victorian designers, its basic silhouette persists as a model for the "Star Wars" blasters seen in current sci-fi films.

Enough eulogizing. The fact remains that in its day (and that day lasted from 1896 to 1937) the '96 Mauser and its variants were popular as military, self defense, and sporting arms. Today it can still turn in creditable groups right alongside some of the more sophisticated space-age self-loaders.

The "Broomhandle" began life on the drawing

boards of engineers in the employ of firearms legend, Peter Paul Mauser. Though Mauser had toyed with target pistols, revolvers, and a manually operated ring-trigger multishot, development of a true Mauser auto pistol was not begun until 1894. Less than a year later, in March 1895, a prototype short-recoil-operated repeater emerged from the toolroom.

Mauser chose the 7.65mm Borchardt cartridge as a basis for the round of his new pistol. In fact the 7.63mm Mauser, as it was to be termed, was dimensionally identical to the round intended for the ill-fated Borchardt, though it was loaded to give its 86-grain bullet a 1,400 fps velocity—some 300 fps faster than its progenitor.

The C/96 ("construction"/1896) as the gun was called, featured a frontally mounted box maga-

This article first appeared in Guns & Ammo

Earliest Broomhandles had reduced magazines and spur hammers. This gun, serial No. 21, was strictly experimental.

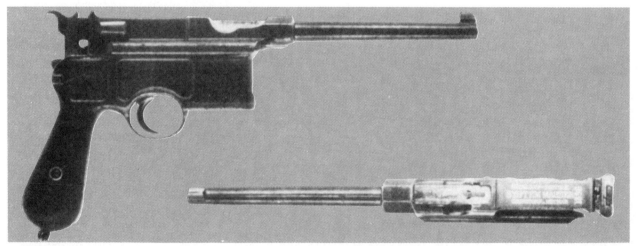

Photos by Jim Brown   *Arms and accessories courtesy of Doug Howser, John Bruchman, Michael Zomber*

Though its design is rooted firmly in the 19th century, the famed C-96 "Broomhandle" Mauser pistol is finding renewed popularity with today's firearms enthusiasts. It was one of the first successful self-loaders on the market. Though its mechanism was somewhat intricate, the C/96 functioned very well and was not prone to breakage. With its shoulder stock/holster, good 300-yard accuracy was not uncommon.

zine that was charged with clips, in the manner of Mauser's highly successful bolt-action rifles.

The grip, which was to remain virtually unchanged throughout the pistol's 40-year production history, was bag-shaped in the manner of many European revolvers of the day. Its ribbed walnut panels and rather stark, tapered appearance accounted for the gun's nickname—the "Broomhandle" appellation remains to the present day.

The C/96 employed a locked-breech, short recoil system involving a rectangular-sectioned bolt which operated within a channel housed by a barrel extension which was one with the 5½-inch barrel. This assembly was slotted and locked within the pistol's frame. When the gun was fired, the bolt (containing an inertial firing pin), the barrel and extension, and locking piece (which resides beneath the bolt, affixed to the barrel extension) recoiled slightly, unlocking the bolt and allowing it to move fully to the rear where it cocks the hammer. A recoil spring (also inside the bolt) returns the bolt to battery, picking up and chambering a cartridge on the way. When the last round was expended, the magazine follower stopped the block from closing, indicating that the arm was empty. The insertion of cartridges (early models came in 6-, 10-, and 20-round versions) and the removal of the clip allowed the bolt to close, and chambered the first round.

Perhaps one of the most intriguing features of the '96 was its almost total lack of screws. It is fieldstripped in much the same manner as a Chinese puzzle.

Making sure the chamber is unloaded and the magazine empty, take a soft-pointed (brass) implement and depress the magazine floorplate button (on the side of the magazine) and slide the floorplate towards the muzzle. With the floorplate forward, the follower and follower spring are in position to be removed.

Next, cock the hammer. Just below the hammer on the frame there is a small, rectangular protrusion—the stripper lever. Using the rim of an empty cartridge, move the lever upward as far as it can go and, at the same time, slowly push the frame forward and the receiver assembly to the rear. The Mauser is now in two main pieces.

During the separation it's possible that the "subframe" (the hammer, sear, and everything that accompanies them) may have come loose and produced another main section. Not to worry. If the subframe didn't come loose from the upper half of the gun, turn the barrel sights down and remove the subframe and the locking block it's attached to.

Insert the tip of a screwdriver in the slot that runs vertically in the rear face of the firing pin. Push in on the pin and turn it 90 degrees to the right and move the screwdriver back and out to

FIELDSTRIPPING

Fieldstripping is fairly simple. **1.** *First check to make sure gun is unloaded.* **2.** *Depress the floorplate button with pointed tool.* **3, 4.** *Slide floorplate forward and remove, along with follower and spring.* **5.** *Cock hammer, push stripping lever up, and push receiver to rear, barrel forward.* **6.** *Separate pistol into two halves.* **7, 8.** *Remove subframe by turning barrel upside down and pulling subframe up, to the rear. The attached locking block will also come free. This completes routine fieldstrip. Bolt and firing-pin assembly can be removed by following directions in accompanying text.*

free the firing pin and its spring.

Now pull back on the lugs of the bolt (note that the bolt is still under spring tension). Keeping the rear of the bolt pointed in a safe direction, locate the bolt retaining button on the right side of the barrel extension and slide it toward the muzzle, pulling it to the side and out of the barrel extension at the same time. The main recoil spring will emerge from the rear of the bolt where it can be removed. The bolt is now free to be pulled clear of the barrel extension.

I know it sounds rather complicated, but once you get the hang of it, the gun virtually falls apart. As an aside, the Mauser is so well engineered that breakage is rare and fieldstripping, for other than a thorough cleaning, is rarely needed. The most vulnerable part on the '96 was apparently the follower spring and a special pouch was stitched to the front of the holster "leather" to contain a spare one for emergencies.

Very early '96s had spurred hammers which soon gave way to round, "cone"-shaped varieties which obscured the sights when forward, indicating to the shooter that the arm was not cocked.

In 1890 these cones evolved into "large rings" which, some five years later, became a smaller, nonobstructive ring.

While the first guns had milled, recessed frames, for a short period guns were manufactured with "flat sides." This was deemed aesthetically unappealing, so the milling was reintroduced and retained throughout the remainder of the gun's life.

Though the first C/96s had no provisions for shoulder stocks, this was soon remedied and a hollow combination stock/holster which slotted into the rear, lower portion of the grip, was offered.

The adjustable rear tangent sight was routinely graduated to 1,000 meters in 100-meter increments and, with the stock affixed, good 300-yard groups were not unusual. In fact, there are reports of telling accuracy out to 600 yards.

Of course Mauser was interested in securing government contracts for his pistol, and though it was tested, official reaction was surprisingly cool.

In 1897 Turkey purchased 1,000 '96s and two

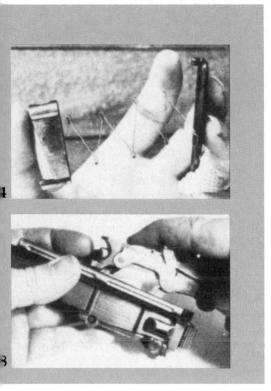

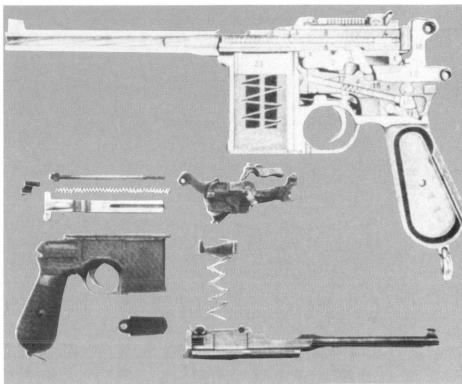

The photo shows the bolt, firing-pin assembly. Drawing: The C/96 Mauser was something of a mechanical marvel with very few screws. The whole could be disassembled in much the same manner as a Chinese puzzle. Breakage of parts was rare.

This is the original "Red 9" Mauser Broomhandle pistol with its shoulder stock/holster and modern 9mm ammunition. The butt of the Mauser's detachable shoulder stock is a hinged lid, released by pressing a button. The hollow wooden stock then becomes a holster that snugly fits the flat-sided pistol. (John Sill photos)

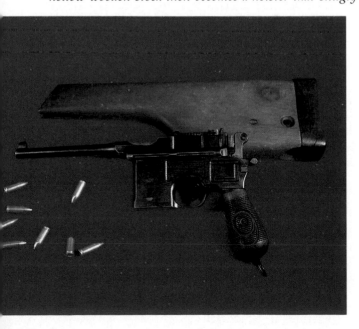

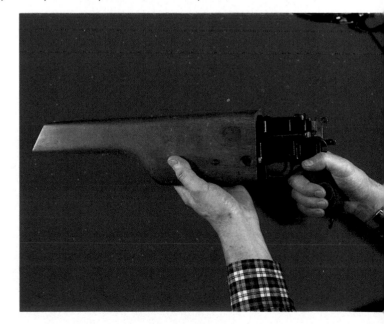

years later the Italian Navy placed an order for 5,000. As well, Siam and Persia used limited numbers of Mausers prior to the Great War.

If war departments were indifferent, however, initially the public was not. The Mauser was a qualified hit with officers of many nations, and not a few were privately purchased and used throughout the globe.

Perhaps the most famous "Broomhandle" afficionado was Winston Churchill who bought one, with a shoulder stock, from Westley Richards & Co. in London, prior to departing for the Sudan where he was attached to the 21st Lancers as a supernumerary lieutenant.

Churchill's account of the combat use of the '96 was one of the earliest reports of the gun's use in battle and surely one of the most vivid and telling.

At daybreak on September 2, 1898, the Anglo-Egyptian Army (including the 21st Lancers) under the command of Horatio Herbert Kitchener, found themselves near the city of Omdurman, only three miles from a force of 60,000 dervish supporters of the Khalifa.

Churchill and seven troopers were sent on an advance patrol, where they found a large number of the enemy awaiting Kitchener's force. They watched the dervishes for some time, eventually riding to within 400 yards of the force where they drew a few stray shots.

Churchill potted at a patrol of Baggara horsemen with his Mauser to no effect. Deciding that discretion was the better part of valor, Churchill returned with his men to the main Anglo-Egyptian force.

Although Kitchener's troops numbered only 20,000, as opposed to the Khalifa's 60,000, the former had more sophisticated small arms, backed by machine guns and artillery. Soon the contestants were battling it out and the 21st was ordered to ride and determine how many of the enemy stood between the British/Egyptian Army and Omdurman. The horsemen rode away in a southerly direction and quickly found themselves within sight of the city. Churchill commanded the second troop from the rear which included about 25 lancers.

The regiment continued to move forward over the hard ground until, about 300 yards away, a large force of dervishes suddenly appeared. The cavalry wheeled into line and sounded the bugle for the charge.

Due to an earlier shoulder injury, Churchill opted to use his Mauser rather than his sword, and prior to the gallop, he returned the edged weapon to its scabbard and drew his pistol. He had practiced with the handgun during the campaign and was confident of its reliability and his own shooting accuracy.

Suddenly he found himself and the regiment racing pell-mell towards the crouched "Fuzzy Wuzzies" who were by now firing volleys at the onrushing horsemen. The two forces clashed and Churchill was surrounded by scores of dervishes, their spears flashing and their rifles booming in his ears. He glanced to the side in time to see a man raise his sword for a ham-stringing cut on his horse. Winston spurred his horse, simultaneously firing two shots at his attacker. Immediately another dervish advanced and Churchill aimed the Mauser and pulled the trigger. He was so close that the gun actually touched the man. Looking up, he saw a chain-mail-clad horseman moving toward him about 10 yards away. Again he fired the Mauser and the rider pulled away.

By this time the 21st was completely enveloped

During World War I, the German War Department ordered some 150,000 C/96s in 9mm Parabellum (left). These guns had large red "9s" on their grips to differentiate them from 7.63s. German Navy also used variations of the Broomhandle such as this 4-inch-barreled 7.63 (right). The gun held 10 shots.

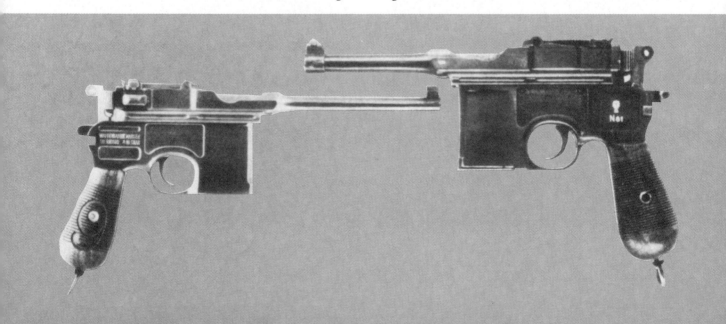

by the enemy and the troopers were fighting for their lives. Churchill found himself cut off from his comrades—everywhere he looked were spears, rifles, and swords. Having no other choice, he hunched down in the saddle and urged his horse forward. The remnants of the 21st were attempting to regroup and he happily headed toward what was left of his troop, about 300 yards distant. Suddenly an unnoticed dervish jumped up in the midst of the British horsemen. Several troopers wounded him with their lances, but the man rose and came at Churchill with his spear. The young subaltern pointed his Mauser and shot the man at a distance of less than a yard.

Though the action of the 21st at Omdurman was one of the few disasters in a successful campaign, Churchill acquitted himself well, abetted by his "Broomhandle." He would later write his mother that the Mauser was, "the best thing in the world."

Churchill subsequently took his '96 to the Boer War, where he also carried it with confidence.

The Mauser continued to undergo minor changes, including experimentation with a proprietary 9mm cartridge which was chambered in a quantity of sporting guns.

In 1912 a new variation of the C/96 emerged.

Though it looked like its predecessors, the grooves in the rifling were reduced from six to four, and the twist itself changed from 1/18 to 1/25 caliber. Perhaps the most salient difference was the fitting of an improved safety (termed by Mauser "NS" or *Neues Sicherung*—new safety).

Now the catch could only be applied if the gun was cocked. Arms with this feature bear a distinctive "NS" stamped on the rear of their hammers.

Again, these pistols continued to be popular with officers and civilians of practically all nationalities (the gun never really did catch on in the United States, though) and many carried them into World War I. Strangely, the British seemed particularly enamored with the "Broomhandle" and there are numerous photos extant of English officers in Flanders with them.

Feeling the pinch of war, the German Government itself decided to adopt the '96 and ordered Mauser to produce 150,000 guns chambered for the service 9mm Parabellum caliber. Outwardly these guns were virtually identical to the 7.63 versions (with the exception of sight graduations), so the grips were incised with a large red "9" to indicate at a glance that the gun was not chambered for the .30 round.

Sporting carbines based on C/96 Mauser were made in limited numbers. They were light and handy but underpowered. Kaiser Wilhelm II used one because of his deformed arm. Bottom photo: Broomhandles and copies were naturals for embellishment, and many were decorated. Most of this work was done after the guns left the factory.

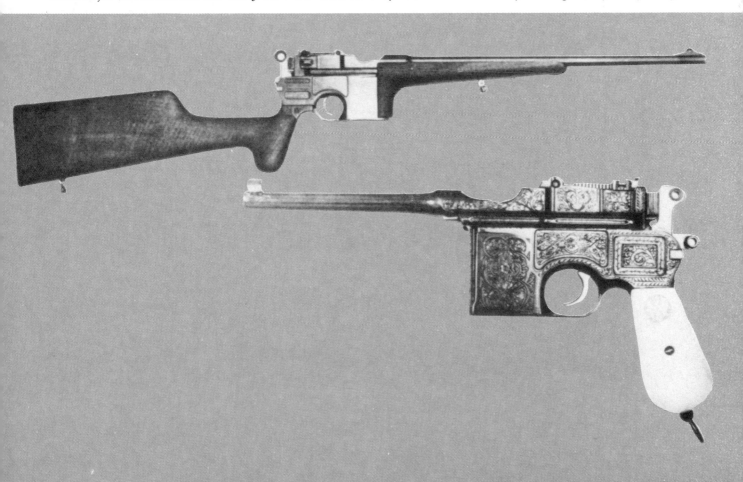

CHRONOLOGY OF C/96 MAUSER

Serial No. Range	Date	Nature of Changes
before 25	1896	—The cone hammer used in place of spur hammer.
50	1896	—"SYSTEM MAUSER" marked on top of the chamber.
before 200	1897	—The locking system changed from one to two lugs. —The barrel contour at the chamber is tapered instead of stepped.
390	1897	—"WAFFENFABRIK MAUSER OBERNDORF A/N" marked on top of the chamber.
975	1897	—The center section of the rear panel on the left side of the frame is not milled out (this feature appears earlier on a few 20-shot pistols). This area is sometimes used for special markings on contract pieces such as the Turkish and Persian.
12,200	1898	—The large ring hammer replaces the cone hammer.
14,999		
21,000	1899	—There is no panel milling on either side of the frame. —A single lug bayonet type mount adopted for retaining the firing pin instead of the dovetail plate. —The trigger is mounted directly to the frame by two integral lugs rather than attached to a removable block. —The position of the serial number moved from the rear of the frame above the stock slot to the left of the chamber.
22,000	1900	—Two integral lugs used to mount the rear sight instead of a pin.
29,000	1902	—Very shallow panels milled into the frame on both sides.*
31,200	1903	—"WAFFENFABRIK MAUSER OBERNDORF A NECKAR" added to the right rear frame panel.*
34,000	1904	—The depth of the frame panel milling increased.*
35,000	1904	—The barrel extension side rails lengthened about a half inch.* —An additional lug for mounting added to the firing pin.* —The hammer changed to the small ring pattern.* —The safety mechanism altered to require that the lever be pushed up to engage it instead of down.* —The center of the safety level knob is no longer milled out.*
38,000	1905	—The short extractor with two ribs replaces the long thin extractor.
100,000	1912	—The rifling changed from four groove to six groove.
130,000		
270,000	1915	—"NS" (Neues Sicherung or New Safety) appears on the back of the hammer. The hammer must be moved back beyond the cocked position to engage the safety.
440,000	1921	—The lanyard ring stud is rotated 90 degrees.
501,000	1923	—The Mauser "banner" appears on the left rear frame panel.
800,000	1930	—The Mauser banner is enlarged. —A step is added to the barrel contour just ahead of the chamber. —The safety is changed to allow the hammer to be dropped from a cocked position, without danger, by pulling the trigger (called Universal Safety). —The front of the grip frame widened to equal the rear part where the stock slot is.
850,000	1932	—"D.R.P.u.A.P." (Deutsches Reich Patenten und Anderes Patenten) added below the inscription on the right rear frame panel.
860,000	1932	—The lettering in the frame inscription is slanted forward.
900,000	1934	—The serial number is moved to the rear of the barrel extension behind the sight. —The two grooves in each side of the barrel extension side rails are elminated.

Early supporter of Mauser was Winston Churchill (here in the Boer War with this Broomhandle beneath right arm).

These nine changes appear out of sequence (either early or late) on three small batches of guns (29,000 to 29,900, 40,000 to 41,000 and 42,600 to 43,900). Most of these pistols are of the "bolo" style, that is they have 3.9-inch barrels, small grips, 6- or 10-shot magazines, and fixed or adjustable rear sights. A few of these pistols show nonstandard barrel contours, barrel extension milling and hammer safety devices. Apparently the factory withheld these numbers from the regular production series and reissued them at later dates.

CHART COURTESY *GUNS OF THE WORLD*, OTTENHEIMER PUBLISHERS, INC.

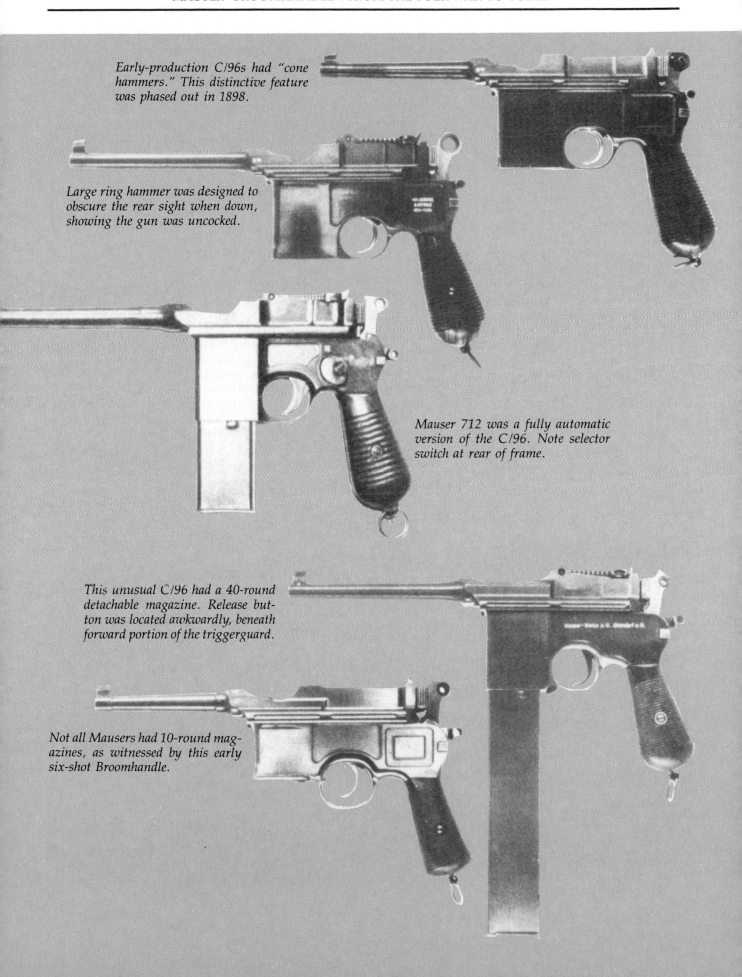

Early-production C/96s had "cone hammers." This distinctive feature was phased out in 1898.

Large ring hammer was designed to obscure the rear sight when down, showing the gun was uncocked.

Mauser 712 was a fully automatic version of the C/96. Note selector switch at rear of frame.

This unusual C/96 had a 40-round detachable magazine. Release button was located awkwardly, beneath forward portion of the triggerguard.

Not all Mausers had 10-round magazines, as witnessed by this early six-shot Broomhandle.

Following the Great War, Germany was restricted, for the most part, from making 9mm pistols by the Treaty of Versailles. As well, barrels on pistols were not to exceed 4 inches. Consequently, Mauser built its '96 in .30 with shorter barrels. Most of those guns featured grips that were somewhat flattened at their bases, in the manner of some earlier models. These pistols were favored by the Russian Communists and were thus nicknamed "Bolos," a period diminutive of "Bolshevik."

In 1930 the safety was changed to allow the hammer to be safely dropped from a cocked position by pulling the trigger. Termed the "Universal Safety," this was to be the last major change on the standard "Broomhandle" before production ceased in 1937. Specialty guns were occasionally devised, however.

Perhaps the most interesting aberration of the '96 was the *Schnellfeurpistole*, a member of a select group of truly automatic pistols.

Originally designed in the early 1930s by Josef Nickl (following the lead of a similar Spanish arm offered by Astra), the M-32, as it was called, had a 10- or 20-round detachable magazine, was recoil operated, and capable of firing 850 7.63mm rounds per minute.

Its silhouette (excepting the extended mag) was virtually identical to that of the C/96; however, it had the added feature of a selector switch on the left side of the frame. When the pointer was placed on "N" the gun could be fired in the normal semi-auto mode. On "R" (*reihenfeur*—repetitive fire), the M-32 could discharge all of its rounds with a single pull of the trigger. The switch was modified in 1936, and the gun redesignated the *Modell 712*.

If imitation is truly the sincerest form of flattery, then Mauser must have been flattered all to blazes. The C/96 was copied (with variations) by such Spanish firms as Astra and Royal and by the Chinese (some in .45 ACP!) in dizzying quantities. In fact Mausers and Astras were particularly popular in China and the Far East. Currently, there has been an influx of these Chinese "Broomhandles" on the U.S. market, creating a renewed interest in the Mauser and its copies. With the relaxation of import regulations there promises to be even more '96s wending their way to these shores.

As an expression of confidence in the burgeoning "Broomhandle" market, ammunition is now being commercially loaded by several firms, and reloading equipment abounds. El Paso Saddlery, Dept. GA, P.O. Box 27194, El Paso, TX 79926, offers a stunning recreation of a Chinese-style leather 96 Mauser holster and copies of the orig-

Left: Despite its front-heavy appearance, the Mauser "Broomhandle" balances very well and is an accurate, pleasant shooter in either 7.63 or 9mm. Right: Ten rounds can be loaded into the C/96 by means of a stripper clip in the manner of a Mauser rifle.

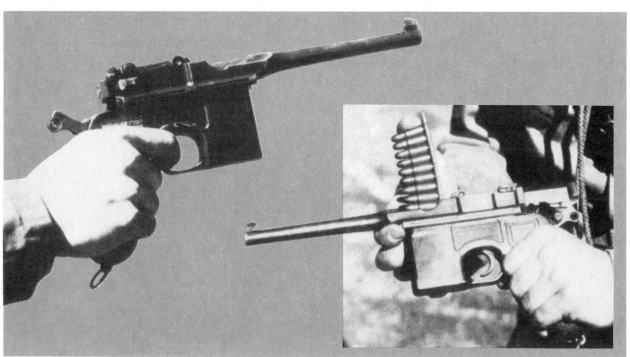

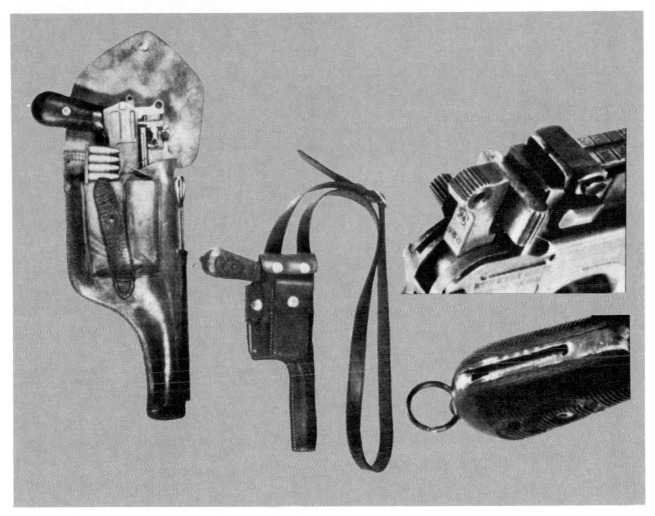

Different types of holsters were officially and unofficially made, including this Nazi-era all-leather model (left). Currently El Paso Saddlery is offering a copy of a Chinese-style holster (middle) in black or brown leather. Guns with the "new safety" (top right) were marked "NS" (Neues Sicherung) to differentiate them from earlier models. Shoulder stock/holsters attached to C/96s by means of slots milled into backstraps (above right).

inal scarce stock/holster leather are being manufactured. New books are coming on the market and reprints of original manuals from Blacksmith Corp., Dept. GA, Southport, CT 06490, and replicas of cleaning tools and copies of composition Bolo grips (Byrons, P.O. Box 796, Dept. GA, Casselberry, FL 32707) have also found their way into the marketplace.

On these pages are represented a few of the fascinating '96 variations, which should help to pique your interest for this fascinating arm even more.

Shoot the "Broomhandle?" I do. Granted one shouldn't take a mint collector's piece to the range, but any good-condition '96 that has been declared serviceable by a competent gunsmith is a joy at the firing line or on a plinking expedition. Despite the gun's front-heavy appearance, it is surprisingly well balanced, and even the antediluvian grip design is not disconcerting when mated to the Mauser's peculiar bodily characteristics.

The Broomhandle is a gun which one can safely predict will never be reproduced. Given today's manufacturing costs, the fitting and machining involved in this intricate, finely crafted arm would probably boost the per-unit cost into the thousands of dollars.

I guess this just reinforces the arm's classicism—a unique product that can never be adequately reproduced. The Mauser Broomhandle is a pistol for its time and ours; it simply transcends technology.

The author wishes to thank Chris Allebe, Doug Howser, John Bruchman, and Michael Zomber for their assistance in the preparation of this story.

Return of the PPK

Pete Dickey

In these days when the firearms industry is experiencing shrinking pains, it is a distinct relief to see a U.S. factory expand.

Such is the case in Gadsden, Alabama, where Ranger Mfg. Co., Inc., is in the throes of adding not one but two new models, the PPK and the TPH, to its line of pistols marketed exclusively by Interarms and marked accordingly.

The present plant's production got underway in 1978 with one item, the Walther American in .380 caliber. The pistol was simply a Walther-licensed direct copy of the PPK/S previously imported from Germany since 1969.

The U.S.-made pistol was, and is, every bit as good as the original, and it had one feature that made it much better: it was much cheaper. The American has prospered and is now available in stainless steel as well as in the original blued version.

Although the American or PPK/S was well received, even its staunchest fans wondered why it was the Walther model selected for U.S. manufacture. Why, they asked, did Interarms/Ranger not make the smaller and more marketable Model PPK? It was a good question, though it is now moot, for the PPK is one of the guns soon to bear the "Made in U.S.A." marking.

If the question must be answered, a brief review of the Walther Polizei Pistole (Police Pistol or PP) line of semiautos seems in order

In 1929, Carl Walther Waffenfabrik of Zella-Mehlis, Germany, produced its first double-action semiauto, the PP. It was small, light, and took eight .32 ACP rounds in its magazine. It was soon made in a seven-round .380 ACP version. Almost immediately, however, a reduced-sized version was introduced that became the PPK (Polizei Pistole Kriminal), its name indicating its suitability for undercover police or detective use. The magazine capacity was reduced by one round in both the .32 and .380 calibers.

The PPK's dimensions went down in overall height, length, and thickness. The weight dropped a couple of ounces, and while both the PP and PPK are here considered pocket pistols, both were commonly used by the German military and police as holster guns.

Many variations of the two models appeared before World War II in relatively limited quantities or as experimentals, including those with scroll or oak leaf engraving, all kinds of plating, aluminum frames, .22 and .25 calibers, a long slide/barrel model, a long grip model, a selective-fire prototype, and the PPK with semishrouded hammer.

Stoeger imported both basic models in .32 and .380 calibers, and in its 1939 catalog offered the PP at $38, the PPK at $34, while the Colt .32/.380 sold at $24. For $2 extra, Stoeger offered both Walthers in .22 caliber, and at premiums of from $5 to $9, all could be had in "blued" or chromed aluminum-framed versions. For $4, the barrel of any version could be had in stainless steel, and for $3 Stoeger would throw in a detachable radium night sight. The Walthers sounded very up to date, but it is doubtful that Stoeger got more than a few examples of even the basic models, for the war intervened and all of Walther's production went toward supporting that effort.

After the war, the French firm Manurhin made the PP and PPK under license from Walther, as Zella-Mehlis was in the East (Soviet) Zone and the Walther firm, showing its usual good sense, rapidly moved to the West Zone. Eventually, a new Walther plant was built in Ulm, Donau, Germany, only a few miles across the border from Manurhin. The French, reportedly using some German raw materials, continued to make the pistols' components which were trucked to Ulm for gauging, quality control, assembly, finishing,

This article first appeared in American Rifleman

Because of its small size, the German-made .380 ACP PPK (shown with white handle) was banned from importation by the Gun Control Act of 1968. It is now made in the U.S. by Ranger Manufacturing and distributed by Interarms. The initials PPK (Polizei Pistole Kriminal) translate "Police Pistol, Criminal." The more-compact TPH (Taschen Pistole, Hahn) translates as "Pocket Pistol, Hammer" and will fire only .22 Long Rifle.

blueing, marking with Walther's name, proofing, final inspection, and to domestic sale or export.

Many PPs and PPKs, plus some offshoot .22 Sporters (single- or double-action on order) came into the U.S. via Thalson, an import firm, first, and then via Interarms, but 1968 saw the last of the PPKs (the most popular) because the Gun Control Act said they were not tall enough. According to the factoring criteria of the GCA, a pistol a fraction of an inch less than 4 inches in height is unimportable—regardless of other factors.

Neither Waltham nor Interarms spent much time in trying to make sense out of that; they simply took the shorter PPK barrel and slide and assembled them to the taller PP frame which gave them, and us, the PPK/S. More logically, it could have been called the PP/S, but the PPK was the more popular parent and its name was the better sales tool.

Some say the "S" stands for "Special"—others

say it's for a comment on the law that bred it— "Silly" or worse.

By 1970, large quantities of the French/German PPK/S pistols were being sold—but only in the U.S. Elsewhere the PPKs continued to sell well.

In 1978, Interarms contracted with Ranger to make the PPK/S here. Walther was glad to give all the help needed, for better sales of the PPK/S would accrue here since duties, Manurhin's profits, and transoceanic freight charges would be eliminated, and since there was no other market for the hybrid pistol. At that time, it would not license U.S. manufacture of the PP or, more importantly, of the PPK.

Today, however, Walther's position may be a bit different. Manurhin is now exporting its own made/marked PPK/S pistols and Ranger's production of the American has finally caught up with demand. More importantly, however, a new 30,000-square-foot Gadsden plant is nearing completion that will be twice as large as the current

Shaded areas on the PPK represent longer slide of the PP and grips of the PP and PPK/S.

plant and will permit multiple production lines to be operated with no loss in production quality. The American, or PPK/S, will continue, side by side with the new guns. Walther will simply get royalties from more than one gun.

Thanks in part to the many souvenirs brought back by returning G.I.s, the PPK will need no introduction on the general market. That is not the case with Interarms' other "new" Walther-licensed pistol. Here, a bit of history is definitely in order, for the *TPH* pistol is new to most American shooters.

It appeared in Europe—and has since been selling well there—just after the GCA '68 came into effect. Those that are now in the U.S. have come in through Interarms only on specific police purchase orders, and the average American citizen, lacking a badge, has never seen one. By GCA '68 standards, it is far too short in height and in length for importation, being 3.6 inches and 5.3 inches, whereas BATF requires a minimum of 4 inches and 6 inches respectively. As in so many cases, what BATF abhors is just what the market adores, but the TPH has more than compactness and an 15-ounce weight going for it.

The initials stand for Taschen Pistole, Hahn (Pocket Pistol, Hammer). Some writers, not all of whom have even seen a TPH, say it is simply a subcompact PPK. That's a good guess, but not precisely correct, as the accompanying pictures will show.

The PPK's features of double-action, quick takedown (by pulling down the trigger guard), fixed barrel, exposed rowel hammer, clean lines and slide-mounted, firing-pin-blocking thumb safety are all present, but the TPH will be made only for the .22 Long Rifle cartridge with a six-shot magazine capacity.

With its PPK/S, PPK, and TPH, then, Interarms will be offering a full line of U.S.-made plinking or pocket pistols suitable for everything from an overcoat to a vest and, despite a half-century tenure, of the most advanced design.

A recent visit to the Alabama factory indicates it will be up to the job.

The common feature of Walther pistols, past and present, German and American, is high quality. The current Ranger plant has proven itself with the PPK/S; indications are the new Alabama plant will follow suit with still greater variety and quantity.

Ranger is a wholly owned subsidiary of the huge Mid-South Industries, Inc., which was founded in 1964 by Jerry Weaver with corporate offices in Gadsden. Its initial success and growth were achieved within the tool and die industry and, with multiple defense contracts, it soon moved into electronics and armaments markets.

Of particular interest to us is Mid-South's Academy of Precision Arts, a technical training school in Gadsden through which each and every Ranger employee must pass. It is equipped with the latest CNC and conventional equipment, and is associated with the Auburn University School of Engineering.

As a result, the people working on the Walthers seem to know what they're doing, as does Chairman of the Board Weaver, whose goal is a southern-based operation eventually making a "full line of best quality firearms."

The Walther, then, may be just a beginning,

WALTHER PISTOL SPECIFICATIONS

Model	PP .380 ACP	PPK .380 ACP	PPK/S .380 ACP	TPH .22 Long Rifle
Overall length	6.68"	6"	6"	5.3"
Height with finger rest magazine	4.75"	4.5"	4.75"	3.66"
Height with standard magazine	4.31"	4.06"	4.31"	—
Barrel length	3.86"	3.34"	3.34"	2.75"
Magazine capacity	7	6	7	6
Weight with empty magazine	23.5 ozs.	21 ozs.	22.5 ozs.	15 ozs.

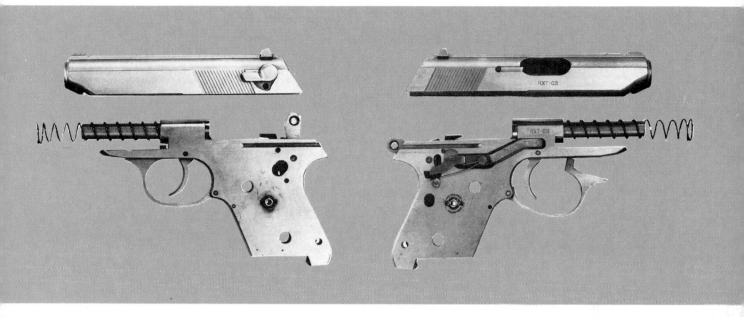

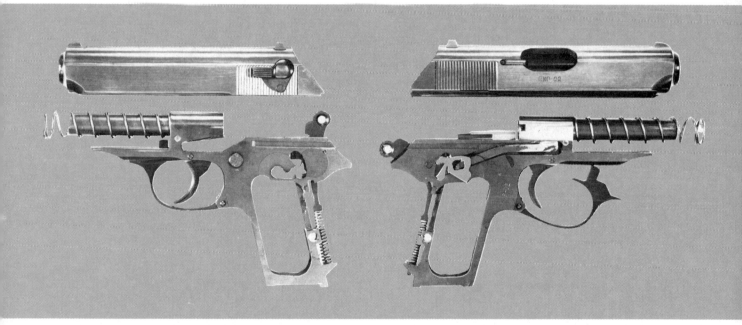

Takedown of the new TPH (top on the row) is same as for the PPK, but the TPH is smaller and simpler, as befits its .22 Long Rifle caliber. Bottom row: Design of the new .380 PPK is the same as that of the original except for the integral spring seat on the rear of the frame.

but the question is when will the TPH and PPK be available.

Ranger's General Manager Bill Quinn and Manufacturing/Engineering Manager Carl Braun agree that the PPKs will begin to be shipped by the end of 1985. Though the prototypes viewed during a plant tour were blued, initial PPK manufacture will be in *stainless steel*.

Stainless-steel TPHs are just now becoming available. An aluminum-framed version, all but identical to the German 11.5-ounce original, will follow.

Interarms' Dick Winter says "available" is a dangerous word. He feels it will be some time before the supply of either model catches up with demand.

For our part, the sooner we get production samples to evaluate, the better we'll like it.

PART THREE

SHOTGUNS

Shotgun Fit:
The Facts and the Fables

Jim Carmichel

When I was growing up in a little town in the hills of Tennessee, there raged an ongoing debate about the breeding habits of possums. Whenever two or more citizens gathered at Toar's Pool Hall, arguments flared over the techniques of the marsupials, and I was invariably awed by the imaginative eyewitness accounts.

After the great possum debate had flamed and smoldered for over a decade, Parnell Purcell, whose dad had sent him to a college up North, showed up at Toar's with a thick book on the begetting methods of wildlife, complete with lurid pictures. "This will fix 'em," Parnell smirked at Toar. "I've got the proof right here, photographs and all."

Very solemnly, the 400-pound Toar closed the book and spread a giant hand across the title. "Son, the plain fact is that *we don't want to know.* Now take your dirty book on out of here." And so it is with the great debate about shotgun fit. The debate is more fun than the facts.

Whenever my sainted predecessor, Jack O'Connor, was particularly displeased by events,

he claimed to suffer from an attack of the "vapors." Though I have yet to succumb to the vapors, I often suffer from similar maladies whenever someone holds forth on the pseudoscience of shotgun fit. For example, after a demonstration of how to determine whether a buttstock is of correct length to the length of one's forearm, I go off my feed for several days and can't keep anything down except champagne and oysters.

Though many shapes and angles contribute to how a stock feels and performs, only five dimensions are traditionally recognized as being of measurable importance. These are length of pull, drop at comb, drop at heel, cast-off (or cast-on), and pitch. If you squint at the fine print in most catalogs, you'll find that only length of pull and drop at comb and heel are usually listed among a shotgun's vital statistics. A manufacturer with illusions of grandeur may also list a couple of other dimensions but, for these, you usually get to pay extra.

This article first appeared in Outdoor Life

Although length of pull is less critical than most gunners realize, the stock must not be so long that its butt rests off the shoulder—against the upper arm. It should mount to your shoulder in a quick, smooth motion, and its comb should be high enough to serve as a platform, or mount, for gun's rear sight: your eye. Excessive drop is a common fault of factory guns. To test fit, close your eyes and snap the gun up snugly to your face; then open your eyes to see if you're properly aligned. (John Sill photo)

Pitch, for instance, is usually measured by putting the butt flat against a floor and positioning the gun so that the rear of the receiver is against the wall. The measurement, in inches, from the muzzle to the wall gives us the degree of pitch-down. Almost all shotguns have pitch-down. Pitch-up exists if the receiver is some distance from the wall when the muzzle is touching it. Pitch-up is almost never found in shotguns. Cast-off simply means that the buttstock is angled slightly to the right of the centerline of the barrel. This permits the shooter to position his eye more squarely behind the rib, which is said to make pointing quicker and easier. Obviously, cast-off is for right-handed shooters. Cast-on does the same thing for left-handed shooters. Most shotguns do not have any cast at all, and about all you have to worry about when buying a gun that does have it is to avoid cast-off if you are left-handed. Don't worry about pitch either. Most American shotguns have a moderate degree of pitch-down, and that's about all there is to that for most shooters.

LENGTH OF PULL

Most wingshooters are more cognizant of and concerned about the length of pull than the other dimensions. This awareness, I believe, is due to the old wives' tales that have been handed down from generation to generation, plus the fact that length of pull can be quickly measured and easily changed. Simply stated, length of pull is the distance from the inner curve of the trigger (front trigger on double-trigger guns) to a central point along the rear edge of the buttplate. In mass-produced guns, the length of pull is about 14 inches but, according to make and model, this may vary quite a bit. Youth-model shotguns have shorter stocks to accommodate shorter arms, and some magnum-chambered waterfowl guns are also shorter than the norm. The reasoning for this is the educated assumption that a waterfowl hunter would be wearing such thick clothing that a full-length stock would be awkward to bring to shoulder.

THE ELBOW TEST

A time-honored way of determining if a shotgun has the correct length of pull is to grasp the grip in shooting fashion and snuggle the buttpad or buttplate in the inside crook of the elbow, with the arm bent at right angles. If the buttplate squares with the crook of the arm, all is supposedly well. If not, the stock is condemned as being too short or too long. I've seen this "test" performed hundreds of times and instant judgments rendered on how well the stocks "fit." The melancholy truth is that the old stock-in-the-elbow trick is utterly useless as a means of determining if a stock is the right length. The only realistic purpose this test could possibly serve is a rough means of determining how long one's forearm is. Otherwise, forget it.

Even if the elbow test were valid, it still wouldn't provide us with very much useful information. Despite the preoccupation of many shooters with the length of their stocks, we can tolerate re-

Left: If the gun fits properly, the shooter's eye will look straight down the rib at the bead or will be slightly above the rib, giving him a view of the top of the barrel. Slightly high comb is better than one that's too low. Right: This gun's comb is slightly too low for the shooter's cheekbone. Because his eye is low, the shooter will tend to raise his head off the stock when firing. (Photos this page by John Sill)

markably wide variations in length of pull without adversely affecting shooting performance. This is evident when we watch shooters of widely varying height, arm, and neck length handle the same shotguns with equal ease. Wingshooters ranging from 5 foot 5 inches to 6 foot 5 inches tend to adapt to the length of the same guns with no problems.

Though a shotgun with a 14-inch length of pull is well suited to shooters of average build, we can tolerate variations of as much as *an inch longer or two inches shorter* and still perform quite well. The main problem with a stock that is too long, other than the obvious awkwardness in mounting or bringing it to shoulder, is that in order to accommodate the excess length, the shooter must hold it somewhat off the shoulder and on the upper arm. This makes the recoil more noticeable or even painful. More often than not, when a petite lady or a youngster complains about recoil, the real problem is that the stock is too long. When the butt is held against the upper arm, rather than the shoulder, the buttplate doesn't make full contact and thus concentrates the recoil in a relatively small area on the edge of the stock. This is why the recoil of even small-gauge shotguns can be uncomfortable.

If you should save your nickles and dimes and have a custom-fitted or "bespoke" shotgun made by a top London gunmaker, you'd probably be surprised to discover that the length of pull is about one-quarter of an inch or three-eighths of an inch longer than you expected. After a few shots, you'd get used to the feel and be quite delighted with your performance.

DROP

Another thing you'd notice about your bespoke shotgun is the way you'd look *over* the barrels when the comb is brought properly to cheek. This

is especially noticeable to shooters who have always used factory-made shotguns, which almost invariably have too much drop.

The comb of a shotgun's stock performs a vital job. It is the platform on which your face rests when you fire a shot. More specifically, it is the mount for the shotgun's rear sight—your eye. If it is too low, your shots hit low. Or, at least, they do in theory. In actual practice, a comb that is too low can cause you to miss in any direction. The reason is that you subconsciously raise your head off a too-low comb in order to see the target. Once you do that, you lose contact with your "sight mount," and the rear sight—your eye—may wander anywhere.

A shooter with high cheekbones needs a rather high comb. When he brings the stock up to his

High-overhead shots (such as skeet Station 8) can be made with fluid swing if stock fits well, but too much length or drop will impede it.

face, his eye is close to the comb line. Conversely, a shooter with low cheekbones and/or a full, fleshy jawline needs more drop at the comb because his eye is supported well above the comb by the shape of his face. Most of today's mass-produced shotguns have so much drop that they fit only shooters who have heads shaped like obese pumpkins. Thus, the majority of wingshooters spend their hunting careers struggling to come to terms with ill-fitting stocks.

When you pick up a shotgun in the comfortable surroundings of a gunshop or your den and snap it to your shoulder, the fit usually seems nice and comfortable, with the eye looking straight down the rib. Even a poorly fitted stock can feel just right in such circumstances because of the way our subconscious aiming mechanism has trained itself to correct for poor stock fit. We can even shoot a nice round of skeet with an ill-fitting stock, further convincing ourselves that the stock is just right. But when the shooting gets faster and the angles are uncertain, as they certainly often are in hunting, our tricky little subconscious gets confused and says: "Aw, what the hell," and we miss shots in unaccountable ways. Follow-up shots are particularly likely to go wide.

To determine if a stock has too much drop at the comb, snap it snugly to the face with your eyes closed. Then open your eyes and see if you are looking over the barrel or at the back of the receiver. Repeat the test several times because you may have unconsciously developed a technique for compensating for poor stock fit. But you say, "What difference does it make so long as I do compensate?"

Alas, excessive drop is a common source of our unaccountable misses. Our compensating system

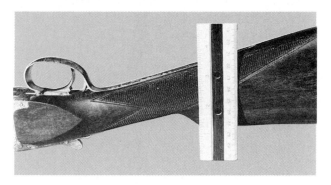

Drop at comb and heel are easily measured by placing gun upside down on table and measuring the space between surface and stock.

breaks down when the going gets rough. Shooters who complain of missing birds that fly at a certain angle may be missing because their method of stock compensation doesn't compensate for that angle.

While we can tolerate remarkably wide variations in stock length, we can tolerate very little variation in comb dimensions without our shooting being affected in some way. This dismal fact brings up the question of which is worse, a comb that is too high or one that's too low.

I'm very much of the opinion that a low comb is worse than one that is a bit too high. In fact, within certain limits, a high comb may even have some advantages.

It's no secret that trapshooters routinely use stocks with extra-high combs. This causes the pattern to hit slightly above the point of aim and thereby provides a built-in vertical lead for the rising component of the target's flight. In a more subtle fashion, a higher comb can offer an advantage with field guns as well. Expert stock fitters at the best English gun houses normally elevate the comb so the shooter's eye is slightly above the rib line, giving him a view of the top of the barrel(s). This tends to put the center of the pattern a bit above the point of aim when tested on the pattern board, but works wonders in the game field.

I'm of the opinion that having the comb high enough so that the shooter is always looking over the barrels is the best way to avoid developing the terrible habit of raising one's head off the stock when firing. This is one of the most common causes of missing. If gunmakers would make their stocks a bit straighter, the problem might not even exist. Like the mother who said she inherited her insanity from her children, raising our heads off the stock may not be a natural tendency at all, but one bequeathed to us by gunmakers who insist on selling us shotguns with too much drop at the comb.

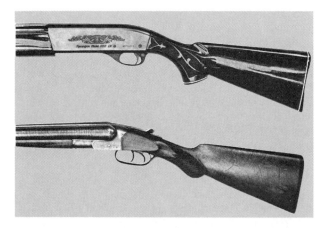

Typical modern shotgun, a Remington 1100, has 1½ inches of drop at comb and 2½ at heel. Baker double, below, has 2 inches at comb and 3½ at heel, typical of older guns. Though modern stocks have less drop, even straighter stocks are needed for the best shooting.

Shotgunning Tips from the Masters

Nick Sisley

The format isn't new. You seek out experts in a specific field, ask them loaded questions, and hope the answers will help the reader learn about the subject. Most of the shotgunners I know are intensely interested in learning more about this arm form we call shotgunning.

Figuring most readers feel the same way, I went to a group of nine specialists in the field, then posed the same 10 questions at each. Their answers, you will find, are most thought-provoking, and, if put into practice, will certainly help anyone's shooting ability.

The questions, and the answers to them, though they come from champion clay-target shooters, should be useful to field shooters.

When you work with other shooters, what is the most common mistake these pupils make? What generally causes that mistake, and how can the shooter's problem be rectified?

Dave Starrett—The common mistakes come in three basic categories. Stance is always a problem. So is foot position, and gun mounting. Poor mounting often is a result of trying to shoot guns that don't fit, or starting out with a poorly fitting gun years back, when the person began shooting. Then the bad habits carry over into later years. The sight of a shotgun is the stock. It determines where the gun shoots for every individual. Once an individual is shown proper foot position and stance, and why they work, he'll generally go along with the instructor. Getting a shotgun stock to fit perfectly requires a person who really knows what he's doing.

Vic Reinders—The most common mistake I encounter is stopping the swing. Often this is caused by the pupil being too tense. I cure it by doing my best to try and make the person feel relaxed. Though we only load one shell at a time, I tell my student, "If you miss this next bird, continue to swing with it, pretending you're shooting a second shot. Pull the trigger again." This suggestion almost always eliminates gun stopping. Too little lead is also a common mistake. When I encounter that problem I'll tell the student, "I don't want you to hit this one. Shoot in front of it."

Ed Scherer—The biggest problem I see is lifting the head when pulling the trigger. This is caused by excessive recoil. The best way to eliminate this problem is to shoot lighter loads or smaller gauges at practice sessions.

Rudy Etchen—By far it's a gun that doesn't fit properly. Generally American gunmakers stock shotguns to fit a 5-foot, 9-inch male who weighs 165 pounds and has a 33-inch sleeve. We can't walk into a store and all buy the same size shoes, coats, pants, whatever, but we are expected to buy the same size shotgun, and live with it. It takes a knowledgeable person or gunsmith to fit a stock properly, and such people unfortunately are few and far between. For almost every shooter, a change in stock length, comb height, and especially pitch will be in order.

Wayne Mayes—I'd say the wrong gun position—like leaning over too far, crouching, putting the head too far over the gun—even shooting with the wrong eye, putting the gun too high or low on the shoulder, out too far on the arm—or too close. Getting help from an experienced shooter

This article first appeared in American Rifleman

THE MASTERS

Dave Starrett has made either the first or second NSSA All-American skeet team every year since 1975, including being named cocaptain of the 1983 team. His four-gun average in skeet has been over 99 for several years. He's well known for his classic form while shooting. An ardent ruffed grouse hunter, Dave is one of the most astute instructors ever to stand behind a pupil.

Vic Reinders has made the All-American trap or skeet teams a total of 21 times, including four as captain. Vic still hunts ducks, but until two years ago, he hunted grouse through the entire three-month Wisconsin season. Noted for shooting a Model 31 Remington pump gun, he recently discovered, through long research, that he'd fired more than 600,000 rounds through it.

Ed Scherer has been on 18 All-American skeet teams, and he has been captain of the All-American senior team for the last three years. Right now, at age 62, he's shooting better skeet than ever. Ed travels all over the country, giving personalized shooting instruction. He's also a life-long hunter who is dedicated to ruffed grouse.

Rudy Etchen made the All-American trap or skeet teams on 16 occasions, including captain four times. In 1980, after 26 years away from the Grand American Trapshoot in Vandalia, Ohio, he was inducted into the ATA Hall of Fame. Rudy shot the program and broke 1,012 16-yard targets without a miss! That means he broke all 600 of the Grand American 16-yard targets, then he went on to win shoot-offs for the AA championship and the Open championship.

Wayne Mayes is unquestionably the greatest skeet shot of this era. For years, Wayne has shot four-gun skeet averages above 99. In 1984, however, he outdid himself, posting a 99.58 four-gun HOA, missing two of 1,450 12-gauge targets, three of 1,300 20-gauge targets, one of 1,200 28-gauge targets, and one of 500 doubles targets. As of this writing, Wayne is over 1,200 straight in the 28-gauge, a new long-run record. Wayne is also an ardent hunter and live pigeon shooter.

Burl Branham has had a shotgun coaching career anyone might envy. Two of his charges at the U.S. Army Marksmanship Unit, Ft. Benning, Georgia, have set world records. Matt Dryke shot the first-ever perfect 200 in world-level skeet competition and last year won a gold medal at the Olympic Games. Dan Carlisle, a bronze medalist at the '84 Olympics, holds the clay pigeon world mark, also at 200, and is one of very few shooters to be a world-class competitor in both clay pigeon and skeet. Branham, curiously enough, spent much of his service career as a high-power rifle coach and competitor, proving that many coaching techniques are applicable to any shooting discipline.

Kay Ohye has made the All-American trap team every year since 1970. He has won virtually every important trap event in the nation. Ohye puts on shooting clinics year round, all over the country, and he has coached an average of nine All-American trap team members each of the last six years. His teaching philosophy centers around making certain students grasp all the basics, but, beyond that, he doesn't insist shooters use his style. He works with bringing out the best in an individual's style and abilities.

Jimmy Prall started shooting when he was only five years old at Lordship, Remington Arms' famous shooting facility in Connecticut. His grandfather built Lordship, and his father managed these skeet and trap grounds. Prall grew up right there at the gun club. By 1957 he was a Junior No. 1 All-American, and he continued to be until he went into the service in 1963. He got out in 1967, and has been an employee of Remington ever since. In the last 12 years, Prall has posted the highest industry high overall skeet average in every year but one.

Dan Bonillas has been a member of the All-American trap team for the last 13 years, and he has been named captain of that team six times. Bonillas is considered to be the greatest shot in trap today. He's also an ardent competition live pigeon shooter.

Dave Starrett

Vic Reinders

Ed Scherer

Rudy Etchen

Wayne Mayes

Burl Branham

Kay Ohye

Jimmy Prall

Dan Bonillas

is the way to learn proper gun position and mounting.

Burl Branham—Stopping the gun is one, a second is lifting the head to get away from recoil, and a third is using the beads to aim at the target. To rectify, first check for dominant eye, then emphasize a good follow-through, explaining how good follow-throughs are important in bowling, golf, baseball, and many other sports. Understanding basic principles goes a long way toward gun-stopping problems. Putting dummy cartridges in the shooter's gun, without him knowing it, can show the person that he *is* lifting his head just before or as he pulls the trigger. The beads, of course, can be removed from the shotgun to prevent using them, thus forcing the shooter to focus on the target. It might take weeks of instruction to eliminate all three of these common problems.

Kay Ohye—Many people, even a lot of good shooters, haven't had proper instruction. They don't know the basics, or proper shooting technique. The biggest overall problem is gun fit, and very few instructors have expertise addressing this facet of shotgunning.

Jimmy Prall—Most new shooters are too anxious. They don't take the time that's required to make sure of the target they're trying to hit. Often they'll lift the head or stop the gun—from being overanxious. Help from another shooter, so that the newcomer gains confidence, is the first step toward overcoming such anxiety.

Dan Bonillas—I think the biggest problem for most is that they aim the shotgun rather than point it. This is probably caused through experience with rifle shooting, though not necessarily. These people have to somehow be made to understand that the shotgun must be "pointed." This is the idea of all the pellets in a shell. Practice with a BB gun, throwing a can in the air, not looking at the sights, and try to hit that can, merely by pointing.

What are some of the ways personalized instruction can aid a shotgunner?

Dave Starrett—Someone who has studied the game is able to see what less experienced shooters are doing wrong. There are certain basics that pertain to shooting, just as there are basics which pertain to any other sport. People new to the game, or those who haven't studied it intensely,

can't be aware of all this knowledge. For example, mismounting the gun, poor foot position, ducking the head rather than raising the gun to the face, rushing the shot, and many, many others—all of these can be extremely hard for anyone but a knowledgeable, experienced shooter, a student of the game, if you will, to pick up.

Vic Reinders—A good instructor can pick up mistakes quickly, plus suggest ways to cure those mistakes.

Ed Scherer—The instructor can certainly tell a person whether he or she is ahead or behind. Many, contrary to common belief, shoot in front of the bird. This is true in the hunting field, too.

Rudy Etchen—Too many instructors try to teach "their" method of shooting, rather than teaching in a way that might work for the pupil. In tournament shooting, Kay Ohye is able to develop very strong pupils because he doesn't teach them to shoot the way he does. He concentrates on bringing out the best in an individual's natural abilities. He also works on the intrinsic value of the person. The hunter who goes to the local gun club hoping to garner good tips won't get them very often, for this is where other shooters are going to show "how they shoot."

Wayne Mayes—An instructor can help a student begin to break targets—fairly quickly. Only when a shooter breaks a target over and over does he begin to know what the sight picture looks like. A lot of shooters break clays and kill birds without knowing how far they were in front or why they hit it. It's important to show shooters why they hit targets, not just why they miss. Only by being able to recall previous sight pictures can a shooter become consistent.

Burl Branham—Everyone can't recognize basic

Though the nine shotgun masters interviewed differ on many points, they universally agree that sound coaching and lots of practice are essentials. Hand-thrown clays are fine for starters.

or not so basic shotgunning problems. Too many companion shooters, even those you might have just met, are self-styled experts in telling you what you're doing wrong. Don't seek instruction from people like this. Too much erroneous information is passed out. Get help from known, respected sources.

Kay Ohye—Instructors help through familiarization with proper basics—correct shooting techniques—hand position, head position, how to move, how to swing, and a lot more. Many who teach, especially at the local club, say, "I do it this way," indicating you should do it the same way. This type of advice can come from a good shooter, and still be incorrect. If a shooter is told, "You shot high and behind," by a team member, that doesn't tell the guy who just missed *why* he shot high and behind. So—personalized instruction is only worthwhile when it comes from someone who knows precisely what he's talking about. Many claim that shooting, especially competitive shooting, is a mental game. That's not right, unless we're talking about the champions at the top of their game. Graduating from an average shooter to a top shooter is physical, not mental. A person has to learn the right shooting basics, the right shooting techniques.

Jimmy Prall—An instructor, simply by standing behind a shooter, can see what problem or problems are taking place, then take corrective action.

Dan Bonillas—Doing so takes advantage of an instructor's experience. In many instances the instructor, at one time, made the same mistakes, faced the same problems, as the hypothetical student we're talking about here. Taking advantage of the services of a good instructor is the shortcut to improved shotgunning performance.

Can trap- and skeetshooting improve one's abilities in the field? How can these clay-bird games help?

Dave Starrett—They absolutely help, mainly because they aid so much with trigger timing, which is important in any shooting, especially in the field. The more you shoot clay targets, the more different angles you're going to shoot at. This experience allows you to better handle subsequent targets—or birds. Also, these clay target games train the shooter to focus the eye on not just the target, but a specific part of it. In field shooting, you want to try, where possible, to zero in on the bird's head. The place to perfect this

technique is through trap and skeet.

Vic Reinders—They definitely help a hunter familiarize himself with his gun. So many put the shotgun away at the end of the season, then never look at it again until the next hunting season. An investment in several boxes of shells, shooting trap or skeet before the hunting season, will pay handsome dividends.

Ed Scherer—They help immeasurably. These games are of particular benefit to hunters if they'll practice skeet or trap for several weeks prior to the hunting season. I've run some tests, talking people into shooting at least eight rounds of skeet in the weeks immediately prior to the opening of dove, quail, woodcock, and grouse season, then have checked back with them at season's end. To a man, these people have shot better, some of them claiming twice the shooting success of previous years.

Rudy Etchen—They help 100 percent. Anybody, novice or All-American, can benefit, especially from skeetshooting. Too many people feel they can put a shotgun away at the end of the season, then pick it up nine months later and do well. You can't do that with a set of golf clubs or ice skates, for example. Why should you be able to do it with a shotgun?

Wayne Mayes—Yes they can, mainly through developing reflex and quick pointing abilities. Field shooters should use a low gun when shooting skeet, concentrating on bringing the gun up the same way and to the same spot—for every shot. By shooting a lot of clay targets, sight pictures are memorized. Then, on a quail or a dove, for example, the shooter knows immediately what the lead should be.

Burl Branham—Yes. They make you a better gun handler. The relationship between shooting a game bird and a clay target isn't the same, but experience in recognizing leads in trap and skeet can be transferred to the field.

Dan Bonillas—I don't know that they're related. I know some good field shots who can't shoot trap well at all—and vice versa.

Kay Ohye—Both these games were started so shooters could improve their scores in the field. There's no other way to improve one's shooting—quickly and cheaply. All the field shots are simulated on the clay-target field. To go hunting somewhere and be able to shoot 25, 50, 100, 150, and 200 shots, like one can every day on clay targets, the cost would be thousands and thousands of dollars. If someone wants to learn how to shoot well, there is no other way.

Jimmy Prall—Both trap and skeet shooting definitely will help—if for no other reason than to give the person greater confidence in gun handling.

If a shooter wants to become a better field shot, what is the most important thing he can do to accomplish that?

Dave Starrett—I think it's to develop the ability to see the bird clearly. Watch the targets you're shooting, as well as all those targets the other members of your practice squad are trying to break. Watch intensely and carefully. Look for the target's spin, the terraces on its dome. Focus on the leading edge.

Vic Reinders—I advise the person to shoot skeet, so he can gain experience with a variety of angles. If he doesn't have a skeet field available, he can set up a variety of shots with a hand trap or a Trius trap.

Ed Scherer—Shoot the shotgun he's going to hunt with on clay targets—several hundred times. Such a person should also have a competent instructor stand behind and help him.

Rudy Etchen—First, get a gun that fits perfectly. In field shooting there's seldom any extra time. It's throw the gun up and pull the trigger. Second, skeet is a big help. So is riverside skeet, hunter's clays, etc. At the Shreveport Gun Club, 2,800 rounds of riverside skeet have been shot, and only five 25 straights have been recorded. Hand-trap targets, extremely hard sidewinders, are also excellent practice to improve one's abilities in the field.

Wayne Mayes—I would say shoot a lot of skeet and trap. When game seasons are closed, get out after blackbirds, crows—keep shooting—a lot. This is the only way anyone can improve at anything—do more and more of it—after learning the correct techniques and basics.

Burl Branham—Shoot a lot, learn to handle guns and learn the techniques of target lead.

Kay Ohye—Get out and shoot. You have to pay your dues.

Jimmy Prall—Practice more at trap and skeet. In field shooting a few people have natural abilities, to the point that they need minimal practice. The average person, however, isn't familiar enough with guns and sight pictures. By handling and shooting a shotgun frequently, a great deal is learned about both.

Dan Bonillas—A hand trap, because so many different shot angles can be simulated, is an excellent place to start, and it's inexpensive. Trap-shooting is more a disciplined angle game, so it's not as helpful in developing the skills that are needed in field shooting.

Talk about the placement of the hand on the fore-end—and why .

Dave Starrett—This can vary, according to the style and shape of the fore-end. The wider models tend to require that you hold back closer to the receiver because the width makes you turn the palm up more. It's difficult to extend your hand too far out on a wide fore-end because the hand will turn too far underneath. On the other hand, a thinner-type forearm, like the schnabel, Fabbri, Beretta and others, allows you to extend the hand out farther on the fore-end comfortably. A good rule of thumb is to get that hand far enough back on the fore-end that you have good freedom of swing both right and left, so that you lock up on your swing through your body's lack of ability to turn any farther, rather than through locking the shoulder, which can be caused by a hand placed too far out on the fore-end.

Vic Reinders—Position that hand where it's comfortable. If you feel like you're stretched, you're too far out. If the hand is too far in, you can't handle the gun as well. For most people the most comfortable spot will be near the rear portion of the fore-end. A hand too far out on the fore-end is more common than one that's in too close.

Ed Scherer—The angle at the elbow, the angle made by the forearm and the upper arm, should be slightly more than 90 degrees—like 100 degrees. A lot of people place the hand too far out on the fore-end. I recommend a loose hold on the fore-end.

Rudy Etchen—This one is more important than most shooters realize. If a person grips the fore-end too tightly he'll start poking instead of swinging. All of the pointing should be done with the right hand (for a right-handed shooter), prefer-

Leaning forward may inhibit the shooter's ability to pick up the target, Dan Bonillas advises, and it may promote head lifting.

The stand-up shooting style is preferred by most of the experts, though they differ on nuances such as foot position and balance.

Should the pointing hand be extended (top) on the forearm, or back near the receiver? The experts disagree but avoid extremes.

ably with a pistol grip that fits. The top of the recoil pad or butt should be at the top of the shoulder, the cheek on the gun. So—it's that right hand, the shoulder and the face that point the gun. The minute you grip the fore-end too tightly you're not going to swing right, whether we're talking about a trap target, a skeet clay, a grouse, a quail, a duck, a dove, whatever. My father had a great comment about successful shooting. "You have to be very careful, in a careless sort of way," he said.

Wayne Mayes—Most of the good shooters will grip the fore-end so the arm isn't completely straight, and so that it's not bent at too tight an angle. Some shooters hold the gun back on the receiver, which makes it more difficult to get the swing started.

Burl Branham—The hand on the fore-end—as you move it toward the muzzle, you can't move the barrels as quickly, but you have better control of the muzzle. As you move that hand toward the receiver you can move the gun faster, but you have less control over the muzzle.

Kay Ohye—This hand must be relaxed. It shouldn't be out too far. If it is, then the farther the shooter has to move his arm to swing from point A to point B. The closer to the receiver one grips the fore-end, the less distance that hand (arm) has to move to swing from point A to point B. However, accuracy of movement is increased by extending one's grip on the fore-end, and vice versa. There's a point in between—where the shooter doesn't feel a strain holding the hand too

far out or too close in. If the elbow of that hand is held too high (level) maximum movement is impossible. If that elbow is tucked way down, like a rifle shooter, it's the same thing, maximum movement isn't possible. there's an in-between position, what I call about a 45-degree angle for the elbow (with the horizon), which most all the top-drawer shooters use—skeet, trap, international and hunters.

Jimmy Prall—I personally place my hand near the rear of the fore-end. This positioning gives me more control.

Dan Bonillas—The fore-end hand can control the gun, and it shouldn't. The barrel or barrels are lighter than the rest of the gun, so the fore-end hand can cause the barrels to "get away" from the shooter. A loose grip up front minimizes this possibility. Gripping the gun tighter at the pistol grip helps with gun control. Also, the farther forward the gun is held with the fore-end hand, the lighter the gun is going to feel, and the easier it's going to be for that hand to overcompensate with control.

Talk about the placement of the hand on the grip—and give us the reason for your theories.

Dave Starrett—I like to have that hand positioned pretty much up and down, rather than along the length of the stock. Doing so gives you more control over the gun's movements, and such a hand position also absorbs more recoil. The main reason for doing this, however, is that most triggers are designed to be pulled directly to the rear, and

69

the type of hand position on the grip that I favor makes this easy to do.

Vic Reinders—Keep the hand on the pistol grip loose.

Ed Scherer—The right thumb (for a right-handed shooter) should be over the top of the grip. The fingers should grip firmly. The forefinger should extend into the triggerguard so that the trigger is contacted at the last joint.

Rudy Etchen—Squeeze the three fingers and thumb as tightly as you can. The reason for doing this is that you don't want to try to pull the trigger with the entire hand. You want only the trigger finger to work. Flinching tends to be much less of a problem when the right thumb and the three fingers other than the trigger finger grip the stock tightly. Also, unless there's a tight hold on the grip, there's a tendency to pull the trigger with the entire hand.

Wayne Mayes—I grip the pistol grip a little tighter than I do the fore-end. I like a snug, though not a firm grip on the fore-end, by the way. I also like to use the pistol-gripping hand to pull the stock fairly tight into my shoulder. With this kind of gun mount, and the snug, not overly tight fore-end, I can swing the gun more efficiently with my right hand and upper body movements. Then the left hand is used less, maybe for up and down movements only.

Burl Branham—The three fingers—middle, ring and little fingers—these are the ones that grip and control the shotgun—at the small of the stock. The thumb and the trigger finger are relaxed. Grip either of these too tightly and chances of not releasing the trigger after firing the first shot are increased.

Kay Ohye—This is the control hand. The grip should be firm, but with the trigger finger loose. Without a firm grip with this hand the shooter will feel more recoil in the face, and he won't be able to move the gun smoothly, or control it as well.

Jimmy Prall—I hold this hand fairly loosely, though I know others tend to use a tighter grip here. Actually, I don't see a significant advantage in holding the grip hand tightly or loosely.

Dan Bonillas—This is the hand that does the controlling of the gun. Don't try to strangle the stock, for this can even make the gun shake. Simply take a firm grip with this hand.

Assuming you favor a stand-up style of shooting, expound upon your reasons.

Dave Starrett—Stand-up shooters can control the swing better because they face fewer variables. The more a shooter leans, the more he has to use additional and different body parts to maintain a level swing, one in which the shoulders don't roll out of level. The more body parts that are moving, the easier it is to make physical mistakes, like rolling the shoulders, what I call looping the barrel, shifting the weight from foot to foot, etc. Stand-up shooters also see the bird better because their heads are erect.

Vic Reinders—Put your feet together, say four to five inches between the heels, then swing comfortably, as far to the left and right as you can, until you feel a strain. Note how far you're able to swing. Then place the feet about 18 to 24 inches apart and try to swing right and left. You'll find you won't be able to swing nearly as far. The freedom to make a wider-swinging arc is one of the main reasons for standing up straight to shoot.

Ed Scherer—It's easier to pivot, easier to have your head swivel rather than sway. If you stand up to track a target, it's very similar to hitting a golf ball. When I was playing golf, the pro instructing me would hold my head in one position throughout my swing—so my head swiveled rather than swayed right or left. The same theory can be applied to shotgunning, and it's easier for a stand-up shooter to do.

Rudy Etchen—Assuming the shooter is right-handed, the right leg is the fence post, your stability with the ground. The left leg, however, should be slightly flexed, then lean the top part of your body slightly forward. Try this. Keep both knees rigid, mount the gun, then have someone swing the barrel muzzle for you. Try the same test with the left knee slightly flexed and you'll see that swinging is so much easier.

All American skeet shooter Ed Scherer checks shotgun's point of impact after watching over his student's shoulder as the student fires at clay targets on the ground at just 15 yards.

Wayne Mayes—An erect stance gives better body control, a bit more tension on the knees and ankles, making it possible to twist or pivot your body with your knee and ankle muscles. You can't twist or pivot correctly unless you're standing up fairly straight.

Burl Branham—We don't teach the erect style at Ft. Benning, at least we don't teach a style that's totally erect, as the Europeans do. With a stance that's too erect the shooter will run out of shooting space sooner than he would if both knees were flexed slightly. In international skeet, the competitors need a great deal of swing—due to the great speed of the target, the low gun position and the delay. By bending the knees slightly and moving with the entire body, more swing is permitted.

Kay Ohye—The only way to shoot is stand-up. What I call stand-up refers to the upper part of the body. Most all the best shooters have gone away from the deep crouch. Movement, better movement, is the reason to stand up straight. If you bend over, you can't turn.

Jimmy Prall—Standing straight up allows easier movements and results in less muscle tension. Both lead to less physical fatigue.

Dan Bonillas—The farther one bends forward, the more likely it is for that shooter to be surprised because he won't see all the targets as quickly. A stand-up-straight shooter finds that the gun comes up naturally and correctly into position each time he mounts it. There's also less chance for head-lifting.

Do you favor swing-through lead?

Dave Starrett—I like to come out with or in front of the bird, but not at the full lead that I really want upon pulling the trigger. Just before pulling the trigger, I'll accelerate my swing. My reason for doing this is that I want to always have my muzzle moving slightly faster than the bird, clay or feathered. I find this to be a very easy-to-control type of swing, especially effective on targets that vary in speed, the type we all encounter regularly in field shooting.

Vic Reinders—I like the sustained lead. A swing-through depends too much upon pulling the trigger at the "exact" moment. If you maintain the right sustained lead for a few feet, it doesn't matter when the gun goes off during that period.

Ed Scherer—The higher the skill of a typical shooter, the more he's apt to use the sustained lead. In the field, some birds require a swing-through, while others are better taken with the sustained lead. I recommend trying both methods during clay-target practice.

Rudy Etchen—To me a swing-through and a sustained lead are the same thing. The important thing is to follow through, not stab the gun at a spot. Whatever creates a good strong follow-through is what should be sought.

Wayne Mayes—I started off with sustained lead. I've tried the swing-through in practice, but I don't think it works as well. Most all the top skeet shooters use sustained lead. There's only a split second in which to break a target if you've swung through, while there's more time with the sustained lead.

Burl Branham—To be a good international skeet shooter, competitors must be able to break the target in three different ways. The primary method is sustained lead. The second method we teach is the pull-away. If you're right on or not far enough in front of the target, the muzzle must be accelerated before pulling the trigger. If the shooter starts behind the bird, he then must be able to swing through it. Learning how to use all three is important, as well as the ability to recognize each. However, the sustained lead is used most often.

Kay Ohye—Sustained lead is common in American skeet, since the gun can be held in front of the target throughout. This isn't so in trap, and in a lot of field shooting. I don't use a swing-through lead at trap, or on most game birds. I try to come through the bird, establish the lead I want, then pull the trigger.

Jimmy Prall—I shoot both—depending upon where I see or meet the target. In skeet I shoot some targets with sustained lead, others with swing-through. In field shooting I also use either, depending upon the individual bird. It would pay all shooters to practice both techniques.

Dan Bonillas—I think swing-through is the best. If erratic targets are encountered, I don't think sustained lead can be as precise. A swing-through shooter fires at a target for what or where it is, not what or where he thinks it might be. This is particularly true at trap, but it also applies to field shooting. A duck, quail, pheasant, whatever, doesn't fly a straight course all the time. When they dip and dive, turn and twist, the swing-through works best for me.

Top-Performance Slugs and Slug Guns

Brook Elliott

In the not too distant past, when the fireside talk turned to deer guns and loads, you'd rarely hear anybody mention shotgun slugs as his favorite round. Rarely hell! It was never mentioned, except by hunters from states where the scattergun and rifled slugs were all they could use. And mostly they just complained about the restriction.

Boy, have things changed! Shotguns designed specifically to throw a single hunk of lead, slugs which are superior ballistically even to some big-game rifle rounds, and a recognition that deer hunting throughout most of the country means shots taken within 70 yards in heavy cover have combined to make shotgun slugs the round of choice for many deer hunters.

Impetus for this changed viewpoint has come from two directions. First, a growing number of states—in whole or in part—now mandate the use of shotguns and slugs for safety reasons. In the past, a serious hunter who thought slugs were inaccurate could always do his hunting in the next state over. This is getting to be less and less true.

These self-same hunters, in a desire to get the most from their gun (whatever it happens to be), have discovered the second point. Neither slugs nor guns that fire them are anywhere near as bad as conventional wisdom would have it.

Take the recently held second annual Shotgun Slug World Championship, for example. In the benchrest category, the top three shooters each came in with groups under 2½ inches while shooting at 75 yards. At other times, in other places, confirmed groups under 2 inches at 100 yards have been fired. There's nothing shabby about that! Hunters who still believe that shotguns and slugs are inaccurate are living in the past.

Of course, this kind of performance is rarely achieved by an off-the-shelf slug gun. Nor, for that matter, do conventional Foster-type rifled slugs produce those kinds of groups—at least not consistently. But, in response to the growing demand for slug guns and ammo that suits the needs of the serious hunter, there's been a growth in specialty slug guns and slugs themselves. Concurrent with that has been a new understanding of how to make even normal guns and loads work better.

Modern-day slug shooting really began in the mid-1950s. Until then, Foster slugs were forced down the bore of a Full-choke, "Long Tom" shotgun. Such a rig may have been good for waterfowling (maybe), but it was the absolute wrong choice for slug shooting.

Recognizing that fact, a band of dedicated slug shooters began customizing their guns to better handle slugs. Starting most often with an Ithaca M37, they'd shorten the barrels and install rifle sights. In effect, they created short, fast-handling .70 caliber rifles, perfect for brush hunting whitetail deer.

No slouches when it comes to recognizing a market, Ithaca began offering a factory configuration of that gun. The M37 Deerslayer featured a ramp front sight and adjustable notch rear sight on a 26-inch improved cylinder barrel. And it took the slug-shooting world by storm. Most other shotgun makers jumped on the bandwagon, and today virtually every scattergun manufacturer offers at least one model in slug configuration.

But Ithaca still leads the way in the attention it gives to its Deerslayers. Most companies do not see slug guns as providing a particularly viable market. The numbers, they feel, just aren't there. So their approach is to take any run-of-the-mill

This article first appeared in Guns & Ammo

More and more slug shooters now consider scopes such as Aimpoint Electronic Mark III (above) and Bushnell Banner Life-Site (below) to be necessary for accurate slug shooting. Target photo: Aimpoint sight helped this 60-yard slug group, but a conventional rig can approximate it.

IC barrel, slap some rifle sights on it, and give it a deer-hunting name. Bore diameters can vary by as much as .030 inch from one barrel to the next.

Ithaca takes a different approach. Each of its barrels is bored to a consistent .729 inch (in 12 gauge) and sights are carefully mounted in line on the barrel plane, just as they are done on rifles. Net result: the most consistently accurate off-the-rack slug guns available.

In addition, the Ithaca guns are the only ones available from the factory that are prepared to accept scopes. All others must be drilled and tapped.

Ithaca also offers a slug configuration in its big Mag-10. With the exception of the H&R single shot, this is the only true 10-gauge slug gun available, and is the most powerful such gun that's legal in the United States.

Real state-of-the-art slug barrels, however, come from the shop of John Tanis, whose Pennsylvania Arms Co. produces both smoothbore and rifled versions. And it's the rifled shotgun barrel that has all serious slug shooters agog. Its accuracy potential, especially when loaded with specialty slugs like the BRI Sabot 500 (pronounced sab-o) is awesome.

With one exception, every shooter at the World Championship used a rifled barrel, usually mounted to an Ithaca M37. My own testing has produced bench-rested groups of 1½ inches at 100 yards.

The PAC barrels are available as blanks, for fitting by your own gunsmith, or in finished form for a limited number of factory shotguns including Remington 870, Ithaca M37, Browning Auto 5, and S&W 1000 and 3000.

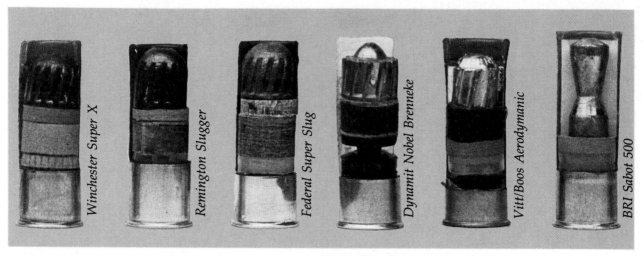

Winchester Super X   *Remington Slugger*   *Federal Super Slug*   *Dynamit Nobel Brenneke*   *Vitt/Boos Aerodymanic*   *BRI Sabot 500*

Though factory-made Foster conventional slugs by Winchester, Remington, and Federal give acceptable accuracy, the three factory specialty slugs at extreme right can be fired with rifle-like precision.

Built specifically to achieve maximum performance with the BRI Sabot 500 slug, each PAC barrel measures .717 inch across the lands, .729 inch across the grooves. Barrel twist is 1-in-34, but other twist rates are available on request. Tanis has found, however, that the 1-in-34 twist is best for stabilizing all slugs. My own experiments indicate he may be right.

Rifled barrels will not perform well, however, with the Brenneke slug, which almost always tumbles when it comes out of the bore.

Originally, all PAC barrels were available with a factory-installed pressure chamber, threaded to accept Cutts Compensator tubes. My original barrel came that way, and I've had no trouble with it. On the contrary, it outperforms any other slug gun I've used, especially when used in conjunction with the Aimpoint sight I've mounted on my Remington 870. PAC found, however, that the chamber would cause slugs to go awry as often as not, so no longer offers it that way.

Being larger than .50 caliber, there is some question as to whether the rifled shotgun barrel is legal. According to Tanis, BATF allows their use because they (1) fire shotgun shells, and (2) are only sold for sporting purposes; they are thus exempt from the rules which define over-.50 caliber as destructive devices. BATF has not formally addressed the question, however. So at this point in time, at least, rifled shotgun barrels are federally legal.

State-level laws are something else. Most state fish and game departments haven't even heard of the barrels, let alone passed judgement on them. In the field, unless a warden happens to look down the barrel, he won't even think to check. Basis for legality, in the absence of specific rulings, seems to be whether the regs state "shotgun" or "smoothbore." Best bet, as always, is to check with your own enforcement people.

If slug guns have come a long way, slugs themselves have come even further. Most of the negatives surrounding slugs no longer exist—if they ever really did.

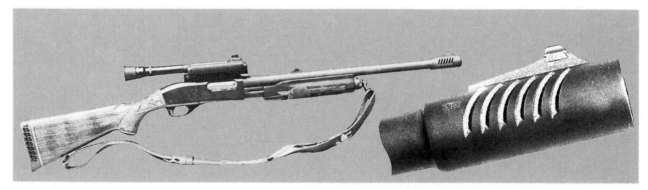

Author's hunting rig (above) has an Aimpoint sight, rifled barrel and pressure chamber (right).

The above demonstration of raw 20-gauge slug power is no revelation to those in the know—shotgun slugs have been performing very well for many years but have now reached a higher performance level.

Blondeau specialty slug from a few years back rearranged this scaled water can in quite a convincing display of brute power.

When Karl Foster patented the rifled slug in 1931, there wasn't too much about it that was truly new. The idea of a weight forward, hollow based projectile had been more than proven with the Minié ball and similar bullets. Foster's main contribution was the thought that rifling the slug would make it more accurate in a smoothbore.

But it doesn't work quite that way. Despite widespread belief to the contrary, the rifling is merely cosmetic and serves no purpose. It is neither deep enough, nor its velocity fast enough, for any spin to be imparted to the slug.

Accuracy from a slug is a function of two things: a good gas seal, and centering in the shotgun bore. Achieve those two and you can shoot as well as with many rifles.

The Foster-type slugs achieve this by having a thin-walled hollow base, which expands into the bore walls, thus effecting a gas seal and self-centering. At least, that's the theory.

Foster's slugs did not accomplish that because molding techniques were not up to snuff, and there would be uneven expansion of the skirt. Often the skirt walls would blow out. A slug that successfully made it to the muzzle faced the constriction of full choking which destroyed any chance of accuracy. So no two slugs hit in the same place twice. The resulting bad press is still with us today. By the same token, the German Brenneke slug, operating on a different principle, was so far superior that its reputation has endured for nearly 100 years.

Today's Foster-type slugs—Federal Super Slugs, Remington Sluggers, and Winchester Super-X—do not suffer the molding faults of the old ones, nor are they often fired from the wrong gun. So their accuracy potential is certainly sufficient for deer at anything under 100 yards. But they are not as good as they can be because the manufacturers must stay aware of possible litigation, and design them to be safe when fired in the wrong gun, like a long-barreled, full choke tube.

Each of these slugs, by the way, is available in one-ounce loadings instead of the original ⅞ ounce. In addition, Federal has a 1¼ ounce load as well.

Groups smaller than 5 inches at 50 yards, or 9 inches at 100 are consistently possible if the slug happens to match the barrel diameter you are using. The speciality slugs, however, are likely to be more accurate from any slug barrel than is any particular Foster slug.

Indeed slug design has, if anything, outpaced barrel design. There are three major specialty slugs available in factory loadings, a few minor ones, and innumerable handloading opportunities.

The Brenneke, in improved version, is still around. Unlike the "rock-in-the-sock" principle that stabilizes Foster slugs, the Brenneke works on a weathervane idea. The gas seal and wad column is attached to the slug with a screw, and this lightweight "tail" always faces away from the wind. Fairly large molded vanes serve to self-center the slug in the bore.

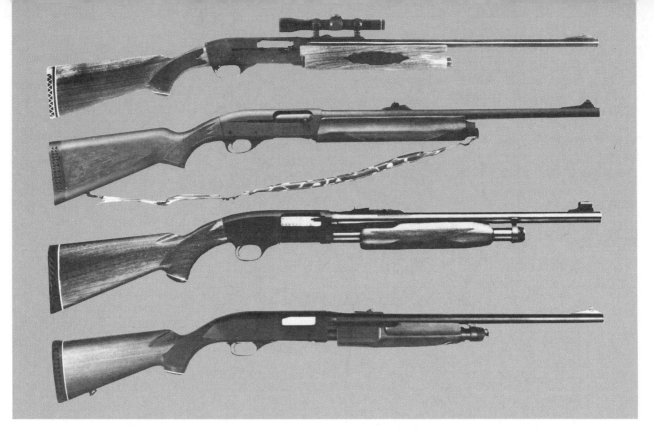

In the manufacture of shotguns genuinely built for slug use, Ithaca led the way with the Model 37 pump Deerslayer and Model 51 auto Deerslayer. Other companies followed suit. Here, from top, are Ithaca 51 Deerslayer, Remington Model 1100 SP Deer Gun, Marlin 120 Slug Gun, and the Winchester 1300 XTR Deer Gun.

Similar in design and principle is the Vitt-Boos Aero Dynamic. At 575 grains, it's the heavyweight of 12-gauge slugs. Superficially resembling the Brenneke, it uses a plastic gas seal instead of cardboard, and the vanes are significantly wider and deeper. Interestingly, the Aero Dynamic, despite its resemblance to Brennekes, does work in a rifled barrel.

Next is the BRI Sabot 500, potentially the most accurate factory slug available. Working on a rad-

Many shotgunners consider use of a sling to be an important part of accurate slug shooting.

ically different idea, the Sabot 500 is a .50 caliber hourglass-shaped bullet encased in a two-piece sabot that brings it out to 12-gauge diameter. Upon exiting the muzzle, the sabot falls away, and the bullet, with its self-stabilizing shape, continues on to the target. Factory testing indicates an astounding 74 percent retained velocity at *200 yards*! And with a decent gun, they should group below 3 inches at 100 yards all day long.

All three of these specialty slugs are available for reloading. In addition, the handloader has more than enough choices to create the perfect load for this gun.

Of late, there is another avenue open. Thanks to Dave Corbin's Hydro Press, at least a dozen customer bullet makers are now set up to produce swaged slugs. A few of them are building and selling their own designs. But all of them are willing to work with slug shooters to produce any desired configuration.

Some of these swaged slugs are something else. Robert Lewis of Power Plus Enterprises, for instance, has developed a custom slug earmarked for police and military use that also has potential for hunting. A finned projectile with a huge internal cavity, it looks like an old blockbuster bomb. Special primers and explosive compounds can be used to fill the cavity for military use. Used "as is" it could be a slug for big and dangerous game.

The right combination of custom shotgun and sighting equipment produced this offhand, 60-yard group. A 440-grain BRI slug was used.

Above left: John Tanis, head of Pennsylvania Arms Co., tries out his slug at a Shotgun Slug World contest. Middle photo: One of five specially made Ithaca M37 guns with PAC barrels takes its place on the line during Shotgun Slug World Championships; the guns were made strictly for this event. Right: Pennsylvania Arms bull barrel and 6 × 40 Swift scope fitted to a Mossberg bolt-action shotgun are preferred by the author at the shooting range. Others use this setup for testing.

Special shapes aside, this could be the ultimate solution to accuracy and powder requirements in that it allows production of a slug that exactly fits the bore dimensions of your gun. Using a .72 caliber, Foster-style slug that he swages himself, for instance, custom bullet maker Bill McBride has very adequately killed Cape buffalo.

The long and short of it is that today's state-of-the-art slugs and slug guns put accuracy and consistent performance well within reach of the hunter who is required to use them. So much so,

in fact, that many hunters now use slugs for both deer and dangerous game.

Slugs and slug guns have come far. In fact, there is now an organization, Slug Shooters International (Box 402, Dept. GA, McHenry, IL 60050), whose membership consists of nimrods using slugs for hunting and competition shooting.

Most of its members have found that there's nothing second class about today's state-of-the-art slugs and guns.

Inside a Patterning Circle

Don Zutz

Although shotgunners casually speak in terms of a 30-inch-diameter pattern, there is a lot more going on inside those circles than the average shooter realizes. Learning the basic percentages of gun/load performance is just a start. After that comes the matter of pellet distribution within said 30-inch-diameter patterning circle, and understanding the characteristics of pellet distribution can be as important, if not often more important, than simply knowing if this or that gun does 75 percent with 8s or 60 percent with magnum 4s.

The first thing a person must realize is that pellets do not spread themselves evenly inside the circle. I have fired hundreds of patterns at various distances, and never have I found a gun/load tandem that would spread its pellets evenly throughout the patterning circle with shot-to-shot uniformity. Such performances are for dreamers, not realists.

Most shotguns, including those with open chokes for relatively close-range shooting, tend to concentrate a lot of their pellets in the center 15- to 20-inch-diameter "core" of the pattern, leaving the outer "annular ring" or "fringe" of the circle irregular. If one were to tabulate the number of pellet hits per square inch of the 30-inch-diameter circle, he would normally find that, at what is considered the basic working range of the gun/choke/load combo, he would have greater density inside the 15- to 20-inch core than in the annular ring.

An explanation for such tendency toward high center densities is easy. Those pellets which aren't deformed by setback pressures or by bore scrubbing or pressures will remain in a rather compact mass in the middle of the shot string. As a rule of thumb, the harder the pellets, the higher the center density for any given gun and load. If one wants to see this carried out to an extreme, one need only pattern with steel loads through a good Full-choke gun. Steel pellets don't deform, and many of the patterns I've shot with them show a maximum working spread of no more than 25 inches at a full 40 yards! Some of my guns will keep 85 to 90 percent of their steel shot inside the 30-inch circle at 40 yards, and practically all of those pellets will be within a 25-inch-diameter ring. Indeed, a studious look at a pattern fired by a Full-choke trapgun with hard shot (high antimony) or with nickel-plated pellets will show just about the same, namely, a very high center density with weak and irregular fringes.

During the years that I have been experimenting with shotguns, I have slowly begun to wonder why we insist on continuing to use a 30-inch-diameter patterning circle; for a careful check of my memory bank and photos proves that the outer 2 to 4 inches of a 30-inch circle are only spottily populated by pellets in virtually every test pattern. The Europeans have used a 75-centimeter circle, which figures to 29½ inches, but that half inch means nothing of significance. We could get a more accurate picture of what a gun/load duo is doing if we went to a smaller patterning circle of, say 27 to 27½ inches in diameter. Most of the pellets striking in those outer 2 to 4 inches of the patterning ring aren't closely packed for multiple hits on target, anyway, and they only add to the overall percentage without contributing much to the actual on-target efficiency. This is the area that cripples game birds and dusts but doesn't break a clay target. One look at the photos on upcoming pages should indicate to the reader just how weak, spotty, and irregular the outer 2 to 4 inches of pattern can be.

I'm sure the shotgun industry won't change

This article first appeared in Shotgun Sports

Traditional American patterning procedure is to determine percentage of pellets striking within 30-inch circle when shot is fired from 40 yards. However, extensive tests show that outer two to four inches forming circle's "fringe" are very spottily hit. For practical purposes, you may learn more about your gun and loads by using smaller circle. You may also want to do some patterning at gun's most common "working range." Inset photo: Shooter examines pattern fired with Full choke at 30 yards, using Old Jake turkey-head target stapled to patterning sheet. He's looking not just for density but for evenness of pellet distribution. Upcoming photos illustrate pattern analysis in detail. (John Sill photos, this page)

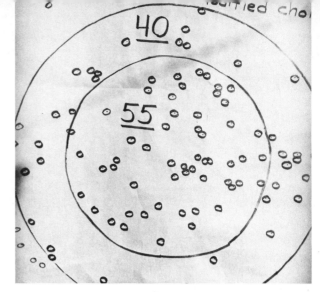

Left: This trap pattern fired at 40 yards indicates the high center density common to trap guns. Note how few pellets populate the outer 2½ to 3 inches of space within the 30-inch circle. For all practical purposes, this pattern's effective spread is 27 to 28 inches, not the full 30 inches. It is typical of all Full-choke guns. Each of the 16 fields has an equal area of about 44 square inches. Photo right: Steel loads throw tight patterns because their pellets don't deform, and the perfect spheres remain in the central mass. Although this is a Modified percentage from a Model 1100 Modified barrel, it shows how tightly steel shot remains in the center of the pattern. A 20-inch-diameter circle encompasses the main distribution here; the outer annular ring has only spotty coverage. This is a common pellet distribution for steel loads.

standard patterning practices because of my observations, of course (although the .410 is in fact tested over a 25-yard range into a 20-inch-diameter circle). But handloaders who experiment widely or who check out their guns with an occasional pattern might well keep my statements in mind. Don't assess the pattern simply by counting pellet holes and figuring a basic percentage. Take a careful look at pellet distribution before jumping to any conclusion.

It is a well-known fact these days that chilled shot deforms more readily than hard (high-antimony) shot, and many shooters tend to believe that chilled shot gives them a patterning advantage by spreading out sooner and, theoretically, filling the fringe. But after reading scores of patterning sheets, I can only say that here is one time when theory and practice do not agree. Patterns that I have shot with chilled and hard shot from the same control load and gun seem to have virtually the same fringe densities, differing mainly in the core densities. This goes back to what I wrote earlier, namely, that deformed pellets split from the main shot mass very quickly, while the healthy round ones remain centrally located in the string. Using various patterning overlays to divide the 30-inch-diameter circle, I find that the fringe of the circle generally has as many pellets from a load of chilled as from a load of hard shot. It is in the 15- to 20-inch-diameter core of the cluster where major differences occur.

Anybody who has patterned a skeet gun at 20 to 21 yards will see much the same thing. Although this choke is supposed to spread its pellets widely—and normally does so, filling about a 30-

to 34-inch area at 20 to 21 yards—further analysis will show a higher concentration of holes in the core than in the fringes or annular ring. Ditto for the Improved Cylinder, which differs from the skeet gun primarily by retaining even more center density.

At one time I owned a Franchi O/U which had 24-inch barrels bored Cylinder and IC. When I patterned with the Cylinder tube, I again found that it, too, showed a definite concentration in the core with an outer ring showing less density and evenness than the average chap would believe. These Cylinder bore patterns were fired at close range, of course, for which type of work it is intended. But even at 20 yards, there was a definite core density.

Some of the most even patterns I have seen came from the Modified choke. As a rule of thumb, the working range of a Modified 12-gauge gun is 25 to 40 yards. It will obviously have a center density imbalance around 25 to 35 yards, but by 35 to 40 yards it will show good balance in many guns. For whatever reason, Modified tends to produce somewhat better core/fringe balance with hard shot than do other degrees of choke within their working clay-target ranges. From what I have seen at the patterning boards, trapshooters using Modified should stick with hard, high-antimony pellets if they wish to derive any advantage from the choke's added spread for 16-yard clays. Easily deformed chilled shot has always left something to be desired in terms of evenness and density in the Modified choke, according to my experiences. The deformed chilled shot merely fly off and leave the already open Modified cluster lacking in fringe density. I have no doubt that many people who complain about weak, chippy hits with a Modified-choke trapgun can lay the blame squarely on chilled shot. A switch to high-antimony shot can improve things, although no true Modified pattern will ever smoke 'em up like a Full choke. You simply can't have your cake and eat it, too! You've got to live with less violent target breaks when using an honest Modified pattern.

I'll often hear a trapshooter or a hunter say, "My old gun sure shoots like a rifle! Either I nail them dead or I miss completely." The idea is, of course, that all the pellets hold in a small area and produce no fringe hits to chip or cripple. Some guns may do that, especially with hard shot or steel. The steel shot now mandatory for waterfowling in some areas is extremely tight-shooting. Steel doesn't deform; hence, the pellets remain in-the-round and hold close to the center of the shot string. This is one reason why many hunters are now missing with steel. Few hunters have the skill to hit with such snug patterns, and the result is they complain because they believe they are hitting and not dropping the birds. The fact is—they're missing with tight patterns, *ultra*-tight patterns.

While trapshooting is far easier than duck shooting at long range, I have no doubt that trapmen are also missing some clays because of too-tight pattern at 30 to 35 yards. Yet, try as I might, I have never found a way to move some core density into the annular ring of a Full-choke pattern short of inserting a so-called "spreader" or "scatter" wad in the shot charge. Using softer shot in the Full-choke doesn't do it; as I mentioned above, the main difference between hard shot and chilled stuff in the Full-choke is found in core density, not fringe distribution.

There are a couple of ways to check your gun/load team for evenness at the normal working range. Shooters may find either of them beneficial in finalizing load selection, because some guns do throw tighter or more even patterns with one load than with some others. It's a crazy world, this one of shotgun performance; but smoothbores tend to be individuals, and trial-and-error testing can sometimes uncover one load that works differently than the others in a given barrel.

One way to check is by using the Thompson and Oberfell method of working a flat disc inside the 30-inch-diameter circle after the shot has been

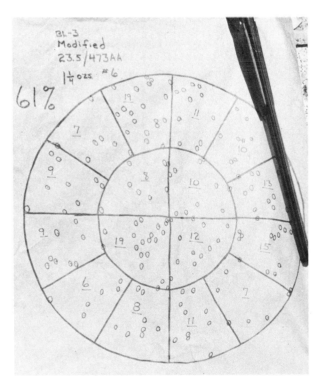
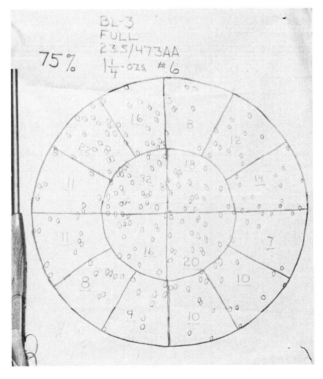

Photo left: Quite often, the main difference between a bona fide Modified pattern and an honest Full-choke pattern at 40 yards is in center density. Here are two examples. This Modified pattern has about the same density in the annular ring as does the Full-choke cluster in the photo at right. The main difference between the two degrees of choke is found in the core, or center density. The FC puts more pellets into the 15-inch core than does the Modified choke at 40 yards. It's worth noting that both show about the same effective hitting area, which is something many shotgunners wouldn't expect.

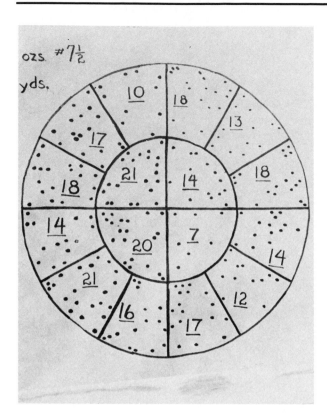
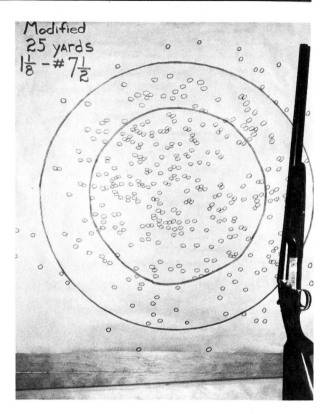

Left: This is the same Modified choke with 1⅛ ounces of 7½s at 40 yards. As noted in the text, pellet distribution within the circle is spotty, even though the numerical count by "fields" is reasonably even in this instance. Thus, learning basic pattern percentages doesn't give one the full picture of a load's performance. Distribution must be eyeballed. Photo right: This pattern was fired from the same Modified choke with the same 1⅛-ounce load of 7½s at 25 yards. Note how it shows a dense center and thin fringe or annular ring, much like a Full choke does at 40 yards. The inner ring is 20 inches in diameter.

fired. According to the T&O tables, an even pattern of No. 7½s should have no more than one pellet-free, 5-inch-diameter patch, on average, per pattern. If a 5-inch-diameter disc can fit more than one place in a pattern without touching or covering one pellet hole, the performance is too spotty or too concentrated, either of which is possible. With No. 8 shot, patterns read at the gun's working range should not have more than one half 5-inch pellet-free patch per pattern. All this assumes a 12-gauge gun using 1⅛ ounces of shot for Full-choke (70 percent) averages. I believe that any casual shooter who finally gets around to working a 5-inch disc through his pattern will be surprised at the weak spots he finds and at the difference between fringe and core distribution.

Another test for evenness is segmenting the patterning circle and making individual pellets count therein. This method isn't perfect, as the numerical results don't mean that pellets are always evenly divided (equidistant). Some areas of the circle may pick up eight tightly-bunched pellet hits, while an adjacent area might pick up seven hits well away from the other eight in the adjoining area. Thus, although seven and eight pellet hits compare favorably in evenness of number, their juxapositions can vary significantly. The best way to use patterning circle divisions, then, is by carefully eyeballing the location of the respective hits. My accompanying photo of a 40-yard pattern fired by 1⅛ ounces of No. 7½s shows reasonable evenness, although the 11 o'clock area has major 'space where the number "10" is printed, and the lower right core quadrant is equally weak. Indeed, this pattern, which is a good representation of a bona fide Modified performance at 40 yards, shows why the Modified trapgun can't smoke clays at handicap ranges: there is no density left in either the core or the annular ring.

An 80-percent pattern sounds good, but it may be too tight for easy scoring. And while a 60-percent cluster sounds wider for more hitting area, the pellet distribution therein may leave weak spots for target slippage. The sophisticated shooter will look to distribution factors, as I've described, in reading his patterns.

PART FOUR

HISTORY AND BLACK-POWDER SHOOTING

Teddy's African Rifles

A. D. Manchester

Conservationist, soldier, sportsman, and statesman, Teddy Roosevelt was the 20th century's version of the Renaissance Man. (Dave LeGate illustration)

East Africa, 1909. Theodore Roosevelt, the All-American President, stalks a Grant's gazelle through a rain squall. It is T.R.'s first day in Africa with a rifle. He carries a sporterized Springfield, possibly the debut of that rifle and the .30–06 cartridge on the still rather dark continent. T.R. is 50 and, at five feet nine and two hundred pounds or more, is in poor hunting condition. Furthermore, his left eye is blind (a 1904 boxing accident) and he can't see well with his right eye. As in the Spanish-American War, he still carries several extra pairs of specs about his person. Whatever Teddy lacks physically, he makes up for with irrepressible enthusiasm and lots of spunk. He lines up the open rear sights, fires— and misses. Teddy has discovered that distances are deceptive in the clear air of the East African highlands.

Never mind. During the next year, T.R. and his son Kermit, then only 19, will take 512 head of game, 296 for Teddy and 216 for Kermit. Some

This article first appeared in Rifle

hunt! Most of the heads and skins went to the Smithsonian Institution.

This was the classic safari. The hunters rode horses. A column of 200 Kikuyu porters balanced loads on their heads. (Natives, it was noted, worked for only a few cents a day, did not require to be packed or diamond-hitched, and could travel faster than animals.) White hunters managed the porters and guided T.R. and Kermit. Each hunter

TEDDY'S ROUTE THROUGH AFRICA

(Saudi Arabia)

Red Sea

Nile

Nile

Khartoum

S U D A N

Jebel (White Nile)

Nile

● **Addis Ababa**

A B Y S S I N I A
(Ethiopa)

Jebel (White Nile)

L. Rudolf (Turkana)

(Uganda)

B R I T I S H   E A S T   A F R I C A
(Kenya)

L. Albert

Kampala

Kisumu

Lake Victoria

(Zaire)

Nairobi

Indian Ocean
(Somali Basin)

G E R M A N   E A S T   A F R I C A

Serengeti Plain

Mombasa

Mt. Kilimanjaro

(Tanzania)

TR and wart hog taken with his Springfield sporter. The rifle was built by personnel at the Springfield Armory and might have been the very first of its breed taken afield by a hunter. Certainly it was the first 06 used in Africa. (Jeff Fitschen map)

Like so many of the animals taken during the lengthy safari, this lioness was destined to be displayed at the Smithsonian.

Even by 1909's standards, the Roosevelt safari was an object of wonder. More than 200 porters were required to carry hunters' and taxidermists' equipment and food.

had a personal gun bearer; each horse had its groom. The likes of such an expedition will never be seen again.

There were other aspects of Teddy's safari that will never be seen again. Imagine yourself on horseback, armed with a Model 95 lever action Winchester chambered for the .30–40, galloping hard after a pride of lions. Neither the conveyance, the rifle, nor the cartridge would find any takers today. Certainly, we wouldn't send our teenage son on such a ride—not unless we're tired of the kid. If this style of hunting sounds like it's next door to suicide, it is probably a sign that the age of the automobile and the magnum have totally overwhelmed us. Ballistically speaking, the .30–40 Krag is a kissing cousin of the still much-touted 7×57 Mauser, which has probably dropped more African game than any other cartridge. Assuming Kermit might have been firing 220-grain government ball ammunition, the muzzle velocity should have been 2,000 fps and the muzzle energy 1,960 foot-pounds. Kermit killed a lioness with one round from his Winchester. Not bad for the old .30 Army cartridge.

Teddy Roosevelt was a president to delight the hearts of American riflemen. He knew his guns, and a brief look at the rifles he chose to carry in Africa should be of interest.

T.R. and Kermit each had a battery of three rifles. Teddy used the Springfield .30–06, a Model 95 .405 Winchester, and a double-barreled .450 Holland & Holland.

The 1903 Springfield needs no introduction to American shooters. The .30–06 is still the most versatile cartridge we have. Rifle and cartridge are part of our national heritage now, but in 1909, both were still new on the shooting scene. In his book *African Game Trails*, T.R. continually referred to his "little" Springfield which fired "tiny" bullets.

In an experiment with '06 military ammo, he put down a rhino. The military round featured a sharp-pointed, 150-grain bullet with a muzzle velocity of 2,700 fps. The first round hit the big bull in the throat. Then began a prolonged chase, during which T.R. occasionally dismounted to fire at the beast's quarters and flank. When the bull wheeled to face his tormenter, Teddy put a round through his chest. The rhino had run four miles, but T.R. was pleased to have killed him. It took nine rounds and Teddy felt they had done their work well. If this sounds like a lot of shooting for just one rhino, it should be noted that on another occasion Teddy needed five full-jacketed bullets from his .405 Winchester and three rounds from his .450 Holland to put down another rhino. The bull rhino he killed with his Springfield weighed over 2,200 pounds. It was an interesting experiment. The .30–06 did "great damage," but Teddy never again used it on a rhino.

Here are the ex-President and his son Kermit with one of their Cape buffalo. The double rifles were gifts from friends and admirers.

T.R. considered the Springfield the "lightest and handiest" of his rifles and with it killed all kinds of American game. Still using military ammo, he killed two giraffes, one with two rounds at 260 yards, another with one round at 410 paces. The military bullets penetrated well, did not split into fragments, and demonstrated great shocking power. Finally, using his Springfield, T.R. killed an elephant with a brain shot. Introduced to shooting during the age of big, slow bullets and black powder, Teddy was brought around to understanding the range of the little modern rifles and the shocking power of their tiny bullets.

Certainly a less handy weapon, and one with a recoil of about 28 foot-pounds, was his 95 in .405 Winchester. The lever-action weighed 8½ pounds and was very effective out to 200 or 250 yards. Launching a 300-grain bullet at 2,200 fps, its muzzle energy was rated at 3,225 foot-pounds. The .405 was Teddy's "medicine gun" for lions. He chose it even when white hunters advised him to use the .450 Holland. The lever-action is fast, almost as fast as a double for a quick, second shot. Moreover, it offers the security of more rounds in reserve. A lever-action would have been a natural choice for almost any American rifleman of the period; Teddy had cut his shooting teeth on lever-actions in the Dakotas.

Using solids, the .405 proved to be strong medicine on Cape buffalo, and with softnoses was none too heavy for some of the larger antelope. T.R. hit one galloping lion at 30 yards with one of the hefty softnoses. The bullet plowed through its flank. A second shot went through the spine and forward through the chest, putting the lion down. The .405 or the Springfield were Teddy's choices whenever he left camp, the .450 Holland & Holland being carried along by a bearer as a backup gun. Considering the weapons available in 1909, the .405 made a good choice, although it is somewhat clumsy to carry if grasped forward of the trigger guard. That was always an objection to the box-magazine rifle, although T.R. does not mention such a problem in his book.

Not a natural choice for most Americans were the .450 Holland & Holland and the Rigby. Magnificent British doubles have never been our cup of tea. At that time, a Holland & Holland could have cost upward of a thousand dollars. Cost aside, the double-barreled rifle has always seemed a somewhat bogus contraption to Americans, as if it were a shotgun masquerading as a rifle. In any event, there are no creatures in North America worthy of their power—and there is nothing bogus about that. The .450 fired a 480-grain slug at 2,150 fps, and delivered 4,930 foot-pounds of muzzle energy. A full-jacketed bullet from a .450 will plow clear through the head of an elephant, head on—a supreme test by anyone's standards. Anything in Africa can be stopped dead in its tracks with two quick barrels from one of those old .450s.

Imagine the recoil! Once, for several days, Kermit had to confine his shooting to a camera. He was that sore from shooting his Rigby. The recoil of such a weapon is considered "objectionable for light men." That's putting it mildly. On one occasion, when T.R. set off his .450, a nearby white hunter developed a nosebleed. Whether or not he'll admit it, with recoil and concussion like that a man doesn't do any more shooting than absolutely necessary. The Holland & Holland was a gift to Roosevelt from a group of friends and well-wishers, and he came to realize that for heavy game—elephant, rhino, buffalo—it was unquestionably the proper weapon. He noted that the bullets from the .405 could break up on massive bones but the heavier .450 bullets tore right through and anchored the animal. One rhino trotted straight at T.R. from a hundred yards, head and tail up. He gave the rhino the right barrel at 60 yards, knocking it right off its feet. As the beast tried to struggle up, Teddy fired his left barrel, knocking it flat again. As the rhino was not quite dead, Teddy reloaded and gave it two more for good measure.

Indeed, Teddy did plenty of shooting. He cheerfully acknowledged lots of misses, probably the result of poor eyesight. Naturally, he felt revulsion at the thought of allowing a wounded animal to escape. As long as a beast remained on its feet, or showed the least intention of struggling to regain its feet, T.R. kept firing. He claimed that during the year-long hunt only two wounded animals got away.

The only other "weapon" Teddy carried to Africa was a gold-mounted rabbit's foot, a good-luck charm sent to him by John L. Sullivan, the ex-heavyweight champion. Just possibly a rabbit's foot was a good idea for the feisty ex-President, a man who would dare to go to Africa with three rifles each using a different action!

Kermit carried two Model 95 Winchesters, a .405, and a .30–40, plus his .450 double Rigby. Kermit at 19 could outwalk, outride, and outrun any African, be they white or black. Before he was 20 he had killed all kinds of dangerous African game: lion, leopard, elephant, buffalo, and rhino. A specialty of Kermit's was absolutely reckless horseback chases after cheetahs. He killed seven, considered an extraordinary accomplishment for the time they were in Africa. He put down several with his .30–40.

The plains of 1909 Africa teemed with immense herds of game. Just as if they were hunting buffalo on the American plains, the two Yanks rode along with stampeding game, firing one-handed and point blank from the saddle. Such sport has its hazards, as when one's mount does a somersault in a patch of thorns. The chases could run five or 10 miles, an exercise guaranteed to toughen up anyone.

Most of the hunting was done in various parts of Kenya. Then the boys crossed Lake Victoria and Uganda for some shooting in the Belgian Congo. Hunting done, they steamed down the Nile to Cairo. It had been an epic hunt. Five hundred and twelve head of game. Today, we can only look back at the expedition with nostalgic wonder.

No American hunter would now venture toward Africa with a .30–40, a .30–06, and a .405. That the old American cartridges proved so effective comes as a surprise to many, so dependent have we become on magnums. Teddy summed it up neatly: "Any good modern rifle is good enough. The determining factor is the man behind the gun." If a rifleman can put his rounds where they belong—and he should study where to put them—almost any reasonably high-powered rifle will do the job.

If you want to see T.R.'s rifles, some of them are in a gun case in an upstairs room at Sagamore Hill, the Roosevelt home on Long Island. Sagamore Hill is one of our national monuments, preserved for us by the National Park Service. Teddy's guns now belong to all Americans.

Black-Powder Shooters Square Off Again at 1,000 Yards

Major Don Holmes

Long-range black-powder target shooting flourished in Great Britain and North America throughout the second half of the 19th century. The legendary Wimbledon and Creedmoor matches epitomized the spirit of that golden age when finely fit and finished single-shot rifles with precise, vernier-scaled iron sights and heavy alloy bullets launched at 1,200 to 1,400 fps were regularly fired in hotly contested, shoulder-to-shoulder, half-mile competition.

Every year, all across Canada, American and Canadian participants in the DCRA's 1,000-yard black-powder program continue to practice that

This article first appeared in Rifle

century-old sport. This article is about that competition, including a brief overview of the century-old DCRA, summarization of the present Canadian black-powder rules, and description of the equipment proven best for half-mile precision shooting.

The Dominion of Canada Rifle Association, Canada's NRA, was created in 1868 from several preexisting provincial rifle associations. Established under its own Victorian Act of Parliament, DCRA has always been assisted by substantial government grants which help operate its various civilian marksmanship and firearm safety training programs.

The present DCRA black-powder program,

This shows Dearborn, Michigan, ASSRA member Ernie Swain firing from the back position during the 1981 Canadian Thousand-Yard Match. Teammate Don Metzler of Goshen, Indiana, is spotting. Note the paper patch in midair, a few feet beyond the muzzle.

Shown with shooting coat and other target-competition gear is 28-inch-barreled Ruger .45–70 rifle with Redfield International front and rear sights. (John Sill photo)

reinstated by popular demand in 1967 (in connection with Canada's centennial year celebrations), is based on more than a century of Canadian long-range international marksmanship experience.

DCRA's 1983 black-powder poster featured a historic woodcut of the 1883 thirteenth annual Wimbledon team shootoffs. The one hundredth Canadian team went to Bisley in 1983. The 1984 poster featured a photograph of the 1982 team match between Canada and the United States. The 1985 poster will feature, by kind permission of the NRA of Great Britain, a photograph of the 1887 Elcho Shield-winning English Eight including W.E. Metford, George C. Gibbs, and the team's Gibbs-Metford match rifles.

Current black-powder rules are contained in Chapter 19 of the official DCRA Rulebook. They endeavor to preserve the flavor and general conditions of the black-powder era. The rules intentionally permit wide, but not unlimited, latitude of individual choice of equipment and shooting styles.

Competitors' rifles may be of period or current manufacture, but must be chambered for authentic, original black-powder cartridges. Although originally a black-powder cartridge, the .303 British round is prohibited and bolt-action rifles of any description are prohibited. Rifles must not weigh more than 15 pounds and must be of a type generally associated with the black-powder era.

The propelling charge in all ammunition fired competitively in the DCRA program must be either black powder, Pyrodex, or a duplex load consisting of not less than 75 percent by weight of black powder. Any cast, swaged, paper-patched, or gas-checked lead alloy bullet may be used. Competitors' ammunition is liable to inspection and inertia disassembly of randomly selected sample rounds on the spot in doubtful or challenged cases.

Any open, aperture or tube sight is permitted, but telescopic sights are allowed only infrequently and when provided for under specified match conditions. Any one- or two-point-attachment military- or civilian-type sling may be used.

Bore cleaning between shots is permitted within the limitations of allocated relay time, and no restrictions other than those already specified above exist with respect to such things as barrel length, stock configuration, weight of trigger pull, rifling style, or twist of personal choice.

Any firing position may be used, including the back position, but unless specifically provided for in given match regulations, no artificial or ground support of any kind is permitted for body, arm, wrist, or rifle.

These rules are the product of a publicly canvassed and reported consensus approaching total unanimity among those who actually participate in the Canadian matches. They have been in effect long enough to demonstrate that they offer neither advantage nor disadvantage to any participating group. As evidence of that, it should be noted that those who use original equipment and straight black-powder loads continue to win their fair share of the prizes offered.

Further acceptance of the universality of the DCRA rules was demonstrated when they were adopted for the 1984 Tacoma Rifle and Revolver Club's two-day 600-yard black-powder Schuetzenfest, and in their adoption by the NRA of Great Britain for Canadian team use in this summer's international match at Bisley.

Choice of equipment can be as simple or as elaborate as each individual's fancy and pocketbook permit. Match-winning combinations are the product of informed selection and attention to important detail. The single-most important notion to get straight from the outset is the crucial difference in performance between mere nostalgia versus serious match-worthy equipment.

Just as open-sighted Winchester .30–30s have no credible place on a long-range .30 caliber firing line, atrocious trigger pulls, slow lock times, and primitive sights, whether on originals or replicas,

At far left, the author; to his right, Diane and Jean-Claude Theriault, Major Edson Warner, and Doug Winger. Two butt boys in front. Jean-Claude and the Martini he's holding are the current Canadian thousand-yard black-powder record holders.

have no hope of success against Sharps-Borchardts, Gibbs-Metfords, or heavy-barreled Douglas-Rugers on the black-powder firing line.

The 1878 Sharps Rifle Company catalog summed it all up by describing its new Sharps-Borchardts, in these terms: "The action has been lightened and the obsolete hammer, that clumsy and dangerous relic of the flintlock era, done away with. Rifles of .45 caliber having proved to give much better results, we have discontinued the manufacture of smaller and larger calibers except on special order."

Gerald Kelver, in *100 Years of Shooters and Gunmakers of Single Shot Rifles*, reminds us that the Best-grade Sharps target rifles proudly featured imported Rigby barrels and were most commonly requested in the .45 2.4-inch chambering, taking a .45–100–550 cartridge, with checkered pistol grip stock, rubber buttplate, and screw-adjustable vernier micrometer sights.

The DCRA long-throat .45–70 Douglas-Ruger incorporates virtually all those specifications, even down to the eight narrow land and wide shallow grooved Douglas barrels of Rigby rifling pattern.

The duplex-loaded .45–70 ammunition typically used in DCRA competition closely replicates original .45–100–550 ballistics. In addition, it offers the additional advantage of using cheap and easily obtainable cases and cleaning-free shooting.

The annual two-day competition, held early in August at the Connaught Ranges 10 miles west of Ottawa, is part of DCRA's larger annual military and civilian matches. All entrants must be DCRA members. The current entry fee in each black-powder match is $10 Canadian, an amount which produces no profit after butt markers' pay, target maintenance, and prize medal costs.

No black-powder matches are fired under 400 yards and all courses of fire, to and including 600 yards, are part of the short-range aggregate. The first day's matches, generally held on Friday of the first full week in August, start after lunch and continue with novelty matches after supper. The novelty events have included 100- and 200-yard Snider service rifle matches fired with century-old issue ammunition, 100-yard black-powder revolver matches, and shoulder-stocked broomhandle and Luger matches.

The more serious Friday afternoon program includes 400- and 600-yard zeroing, team shakedown and selection events, and practice shoots. Hot competition begins Saturday morning with 400- and 600-yard individual and three-member team aggregate events for the Loyalist Cup. The course of fire in all matches is as many rounds as it takes to find paper, then two convertible sighters, followed by 10 rounds for score at each distance. Convertible sighters are declared for score

The Challenge of the Single-Shot Rifle

Ruger advertising correctly identifies the Number 1 single shot action with the authentic traditions of the classic Wimbledon and Creedmoor eras.

before starting the record string, in which case the marker deletes the last one or two unfired rounds from the competitor's score card.

Shooters are normally squadded two to a target. They usually, but not necessarily, fire alternately. Each marks the other's score card.

On Saturday afternoon, the 700- and 900-meter aggregate is fired, both for individual score and in previously declared team competition for the Ranger Cup. The distances currently used are metric because Connaught is in the midst of conversion to the NATO standard.

Half-mile-plus black-powder shooting with multisecond time of flight and high trajectory requires considerably keener wind doping skill than is necessary in .30 caliber competition. Shot to shot adjustments are required before each round is fired to stay on a distant target which subtends less than four minutes from center to edge in the vertical plane and six minutes in the horizontal. The 40-inch-diameter aiming mark the Canadian team will face at 1,200 yards at Bisley almost subtends a .025-inch front sight aperture for which a .045-inch ring is generally used, depending on lighting conditions.

Sadly, the North American market for precision iron sights is drying up to the extent that there is no longer sufficient demand to maintain economic domestic manufacture. World-class Canadian and Bisley shooters now use either the Australian Central, British Parker-Hale, or Ger-

man Anschutz micrometer peep sights. All three are designed for modern .30 caliber shooting which needs less than half the elevation adjustment range required by long-range black-powder competition, and which generates milder recoil than is encountered when launching bullets weighing as much as 650 grains.

The Central sight is the lightest and sturdiest of the three. With its powerful lever-locking thumbscrew, it can be snugged up tight enough to resist heavy recoil.

The solution to elevation extension is to mount two separate bases, one above the other, so that when the sight's lower base elevation range runs out around 600 to 800 yards, the sight can merely be returned to zero elevation and reattached to the upper sight base for ranges between 600 and 1,200 yards.

Black-powder competitors, firing at targets beyond 800 yards, are typically able to shoot, ground their rifles, roll over, and leisurely gaze through their spotting scopes to await the telltale puff of dust in the distant backstop announcing their bullet's safe arrival. A squadded pair with a 1/100-second digital stopwatch can time bullet flight with surprising shot-to-shot consistency. Human reaction time is essentially the same at both ends of the sequence, as clicking the watch on and off as quickly as possible can easily demonstrate. In fact, it's been shown that most individuals' response time, even after minimal practice, seldom exceeds a few hundredths of a second.

Canadian black-powder team members Eric Greer, left, and Bryan Kaufman won 21 of the 39 medals offered at the out-to-600-yard Statue of Liberty Centennial Match held at Fort Dix, N.J., in April 1986. Greer additionally won the cased Navy Arms Rigby rifle replica Grand Aggregate. Kaufman additionally won the 400 yard single offhand shot Hero Match and hand-painted Schuetzen target as a memento.

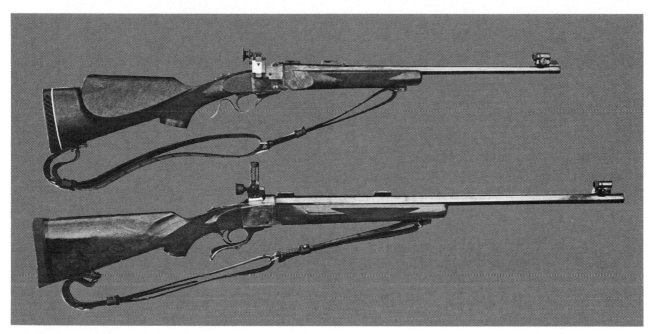

Typical .45–70 black-powder match rifles. Top, this standard Ruger No. 3 with its factory barrel won the 1975 long range aggregate. Note the left-handed Parker-Hale sights. Below, a heavy Douglas-barreled Ruger No. 1 with long-range Central sights and extra-thick buttpad. An identical rifle beat all comers at 800, 900, and 1,000 yards at the 1984 matches.

Chronographed muzzle velocity, long-range time of flight (to a few hundredths of a second), and true long-range elevation taken by careful use of that invaluable artillery instrument, the gunner's quadrant, produce better ballistic coefficient and range table data for cast bullets than is commonly available for most modern smokeless ammunition.

A gunner's quadrant is the artilleryman's caliper. When used with a black-powder rifle, the rifle must first be leveled on sandbags and pointed directly at the distant target. The angle can be measured with the quadrant oriented along the barrel, the scope mounting blocks or some other convenient part of the rifle. Then the sights are raised until the bullets are actually hitting the target. A second angle is measured, using the same basis as the first. The difference between the two angles is the true elevation for the range and the load. By measuring the angles at all the ranges fired, a workable range table for a particular rifle and load can be made up making future sight adjustments less traumatic.

My own World War I Canadian artillery quadrant, shown in an accompanying photo, is accurate to 20 arc seconds, or approximately three inches of elevation at 1,000 yards.

Team shooting is the culmination of the black-powder program experience. An individual shooter who messes up early in a string, has only himself to consider. Team shooters have a shared

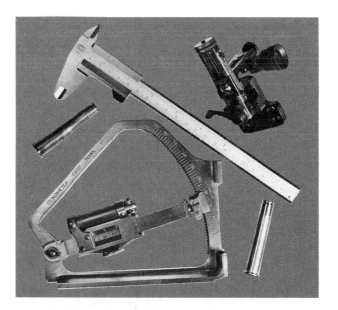

The World War I British Gunner's quadrant, at bottom of the photo, is accurate to 20 arc seconds. The Australian Central Sight, in the upper right-hand corner, offers 50 minutes' windage adjustment, right and left, plus an eyepiece group that includes a colored filter and a place to fit a prescription lens. The elevation scale has 7.62 NATO calibrations as well as the usual graduations in minutes. The empty .38–56 and .45–70 cases are left unprimed and are employed when bore sighting. The instruments shown will enable a shooter to develop range tables superior to those for modern rounds.

responsibility. To be competitive, they must be prepared and able to coach one another, should use similar equipment, share sighting corrections, and persevere even in personal adversity because every point counts. Besides, the other teams may be experiencing similar difficulties. While teams of eight are the international standard, DCRA domestic matches permit teams of three for most matches. Teams of two are too easy, while more than three have been found to be too difficult for many isolated groups of shooters to organize.

THE BLACK-POWDER MATCH RIFLE

This summary is offered for those interested in building a competition-worthy black-powder match rifle.

Original actions, if of sound design and in good condition, may be more expensive than equally suitable actions of more recent manufacture. Additionally, old actions are always liable to parts breakage and replacement.

Actions with large outside hammers, modern or original, cannot deliver the crisp trigger pulls and desirably fast lock times of such better actions as the Sharps-Borchardts, Biggs-Farquharson, or the modern Ruger No. 1s and 3s. Many otherwise desirable inside striker actions, including the Martini-Henry and various Peabody variants, limit loaded cartridge length and preclude cleaning from the breech.

The most frequently seen and most successful black-powder match rifles have been based on the Ruger No. 1 falling block action. The least expensive solution is to find a good used No. 1 with the beavertail fore-end, and to rebarrel. Ruger factory woodwork is better value for the money than any equivalent amount spent on custom stocking.

The caliber of greatest economy for 1,000-yard black-powder target shooting is .45–70. Cases are the largest of their kind still in production.

Although many small makers offer broach-cut barrels in .45 caliber, few have the sophisticated quality-control instrumentation used by the larger and better equipped barrel makers. Douglas offers premium grade, chrome moly, double heat treated, button-rifled barrel blanks in .458 in a wide variety of twists, lengths, and outside diameters.

The steel fore-end extension of the Ruger single-shot action precludes barrels with a breech diameter exceeding 1.150 inches. Douglas Number 7 barrel blanks are straight taper from 1.150 inches at the breech to .875 inch at the muzzle. Douglas Number 8 blanks go from 1.150 at the breech to 1.00 at the muzzle. Such blanks are normally delivered 26 or 27 inches long. Thirty-

inch blanks can be had on special order at extra charge.

Twist is a matter of personal choice. Barrels with 14-inch rates perform superbly. The standard .45 Sharps, Rigby, and Winchester twist was 18 inches. Douglas also offers a 16-inch version.

A Ruger No. 1 with a Douglas Number 7 contour barrel finished to 26 inches weighs approximately 10 pounds. With a 30-inch Number 8 contour barrel, weight rises to approximately 12 pounds.

Not much less than 150 minutes, or approximately 1½ inches of elevation adjustment, are required to a range out to 1,000 yards in black-powder .45–70 shooting. Primitive unscrew-the-eyepiece, adjust, retighten-the-eyepiece, rear

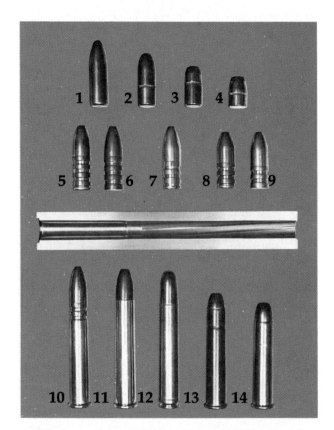

Jacketed .45 caliber bullets: (1) 600, (2) 500, (3) 400, and (4) 300 grains, respectively. Second row, Hoch nose-poured cast bullets: (5) 600-grain gas-checked and (6) plain-based versions; (7) optimum .45–70 long-range bullet, a 580-grain gas-checked design; (8) 540-grain gas-checked and (9) plain-based styles, the minimum recommended weight for long-range shooting. Below them, .45–70 chamber section, showing the long throat favored by Canadian competitors. At bottom (10) a .45–70 with 580-grain bullet seated out to fit the long-throated chamber; (11) a .45–110–550 Creedmoor-type round; (12) a .458 Winchester Magnum; (13) a .45–90 (2.4-inch case) and (14) a .45–70 (2.1-inch case).

sighting arrangements are hopelessly inadequate for serious long-range target shooting. Heavy Redfield Olympic, Parker-Hale, and Anschutz sights quickly shoot loose at the high extensions and jarring recoil levels involved. The best solution to the sighting problem has been to mount the light, sturdy adjustable Australian Central rear sight on a two-height rear sight base.

Possibly no other single factor is as conducive to a rifle's accuracy potential as the quality of its chambering. Many of the best black-powder match rifles have been painstakingly chambered using custom Clymer reamers featuring long, straight cylindrical sections, exact groove diameter throating shanks, and removable press-fit bore diameter pilots.

Headspacing is held to five thousandths *less* than the nominal .07-inch .45–70 minimum. Up to five percent of some lots of factory brass must occasionally be set aside because of overly thick rims which cannot be chambered. Fired brass from such chambers falls freely in and out of standard-tolerance .45–70 chambers.

Brass expands only a few thousandths in such chambers and bullets fit neatly in fired cases. One or two grease grooves remain exposed on bullets expressly loaded for such chambering. Case powder capacity is maximized and the bullet snugs precisely into the long cylindrical throat section before firing. Judicious adjustment of overall car-tridge length results in mating the bullet's first driving band with the rifling during the final camming motion of the Ruger's breech closure.

Another common refinement frequently seen on better match Rugers is the addition of spacers and/or a thicker hard rubber buttplate to extend the stock's length of pull for prone shooting.

MATCH AMMUNITION

Most old hands know the following by heart. These are the easy and simple tricks of producing superior .45–70 duplex loads for the 1,000-yard program.

Cases: Remington and Federal are the only brands which come, if ordered as empty unprimed brass, uncannelured. Remington boxes accommodate extra-long handloads; Federal do not. Cases should never be resized more than minimally and case mouths should be left bellied outward for concentric friction fit in the rifle chamber.

Duplex Loads: DuPont SR-4759 is made for uniform low-pressure combustion and is the duplexing powder of optimum choice. Hercules 2400 can be used as an acceptable alternative if that becomes necessary.

Curtis's & Harvey's (left) was universally acknowledged as the best black powder available anywhere throughout the original era. Modern Curtis's & Harvey's (right) is made for the parent company outside the U.K. and is not an equivalent quality product. Middle: Gearhardt-Owen GOEX brand black powder is a superior product and the best currently available.

DUPLEX LOAD CHARGES		
(grains)		
SR-4759	Fg or FFg	black-powder equivalent
14	42	84
15	45	90
16	48	96
17	51	102
(Smokeless charge in first, followed by black powder)		

Black Powder: Gearhart-Owen GOEX Fg black powder gives very uniform results. FFg may also be used. The most consistently accurate ammunition has the powder weighed, drop-tubed into the case, and never crushed during bullet seating.

Primer: No clear difference between primer types has been documented in long-range black-powder shooting. Magnum primers produce an average of 20 fps more muzzle velocity than standard Large Rifle primers with the same duplex loads.

Bullets: Provided in all cases that rifling twists of one turn in 20 inches or faster are used, bullets in the 400-grain weight range, well cast, of sound design, which provide a good fit with the chamber throat and bore, perform better than heavier bullets do out to 400 yards. Between 400 and 600 yards, bullets of all weights perform equally. Beyond 600 yards, bullets in the 540- to 580-grain range excel. Nose-pour moulds produce a higher proportion of good castings. Gas checked designs shoot cleaner, lead less, and seem just as accurate as equivalent plain based designs.

Bullets should be seated as far out of the case as practicable and seated without undue pressure in the chamber throat.

Bullet lubricant should be moist and plentiful, especially in hot weather. Half & half beeswax and Alox 2138 has proved an eminently satisfactory lubricant.

Bullet bearing bands should be groove diameter; the shank, bore diameter less no more than one or two thousandths just ahead of the bearing bands. Bullets should be seated far enough out of the case to touch the rifling on breech closure.

Hard wheelweight or linotype bullets alloys work best, and crimp-on Hornady style gas checks are most popular with those who have experimented with the different types.

DCRA rules require not less than 75 percent by weight of black powder or Pyrodex in all ammunition fired competitively in the 1,000-yard program. It is statistically known that each grain of smokeless powder is equivalent to three grains

Page 95 of the official history of the NRA, Americans and Their Guns, *records that 40 percent smokeless proportion duplex loads using Brackett's Sporting Powder were being used by the American national team in international competition at Creedmoor itself at least as early as 1882.*

of black powder in a duplex load. For the .45–70, depending upon bullet weight and seating depth, the charges shown in the table constitute drop-tubed, uncompressed full-capacity loads.

The smokeless charges go in the cases first, of course, then the black powder or Pyrodex. Those loads shoot clean. Bore cleaning isn't even required between strings. In hot, humid, or rainy weather, rusting will begin within two hours. Cleaning between matches or over the lunch hour is therefore appropriate under those conditions. Fired cases need not be cleaned the same day except when similarly humid conditions prevail, but whenever deprimed, they should be soaked in a hot baking soda and water solution to neutralize salts and pyro acids.

All bullets should be visually inspected for surface imperfections *and individually weighed* before being accepted. Second-string bullets should be set aside for practice or sent back to the melting pot.

Experience of the Connaught matches can be of unexpected value to American shooters. The pace is leisurely, the food excellent, the exchange-rate favorable, and the friendliness and courtesy of the shooters unbelievable. Plan on visiting Ottawa this summer!

Sir Charles Ross: His Controversial Rifles and Cartridges

H. V. Stent

Praised, damned, and fiercely debated for nearly 80 years, it must be the most controversial rifle-cartridge combination ever manufactured commercially in North America.

Some still swear it was the slickest, quickest, strongest bolt-action ever made. Others swore at it as a killer at both ends. Adherents hailed it as the ultimate in accuracy. Critics doubted it would hit a house beyond 40 rods. Hunters claimed it killed like a thunderbolt, but whispers of failure always dogged it.

Amid the welter of conflicting arguments, one cold fact stands out: so far ahead of its day was the .280 Ross that as of now, 65 years after its demise, major arms makers copy its action, and duplicates of its cartridge rank high in hunters' favor.

Look at a current Remington pump or auto rifle, including the newest Models Four and Six. Look at a Browning BAR or BLR, or a Winchester Model 88 or 100 circa 1955–1975. Within their steel breeches you will find the old Ross idea of a "bolt within a bolt," the outer one sliding straight and by means of spiral ribs and grooves making the inner one rotate to lock or unlock its lugged head. Whether the actions are actuated by gas (self-loading), pump handle, or lever, the principle is the same as the manually operated Ross.

As for its .280 cartridge, what are many of the current big-game favorites but .280s? Indeed, Remington calls one of theirs a .280, but it's the same .30–06 case necked down to take .28-inch bullets. Bigger and still more popular is the Remington 7mm Magnum, introduced in 1962. Its 150-grain bullet at 3,110 fps and 175 grains at 2,860 fps are considered by many to be ideal for soft-skinned big game.

Ross experimentally necked a .30–06 down to .28 way back in 1906 when it first appeared. In 1907 he developed his larger .280, which drove a

This article first appeared in Rifle

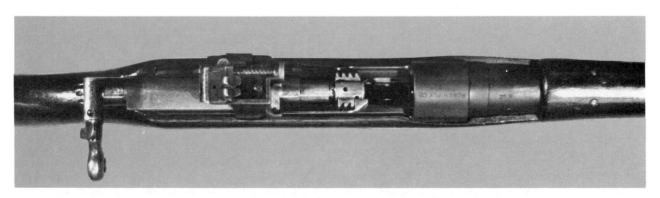

Overhead view of .280 Ross. The unique threaded locking lugs were given mixed reviews by shooters. The receiver sight was a marvel, offering a choice of either open notch or adjustable aperture.

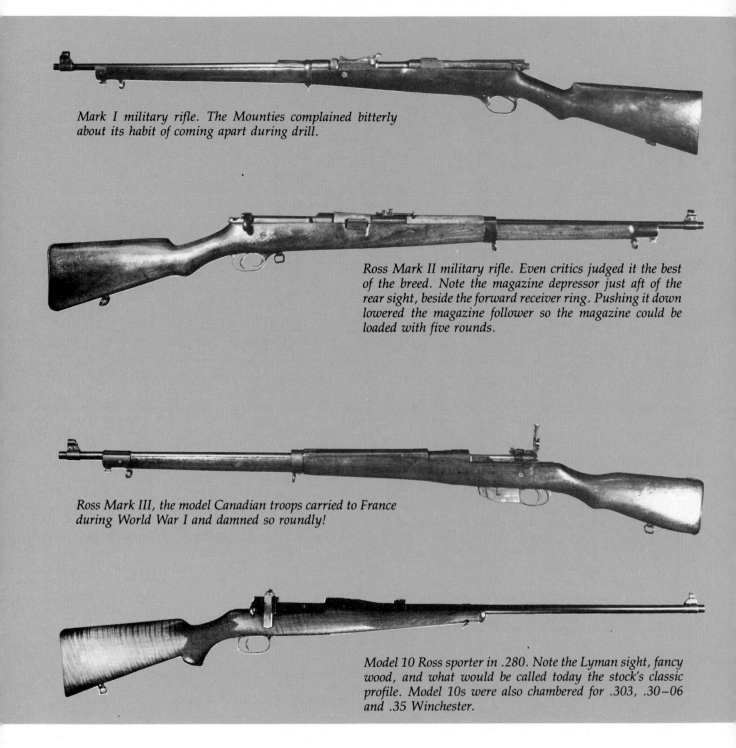

Mark I military rifle. The Mounties complained bitterly about its habit of coming apart during drill.

Ross Mark II military rifle. Even critics judged it the best of the breed. Note the magazine depressor just aft of the rear sight, beside the forward receiver ring. Pushing it down lowered the magazine follower so the magazine could be loaded with five rounds.

Ross Mark III, the model Canadian troops carried to France during World War I and damned so roundly!

Model 10 Ross sporter in .280. Note the Lyman sight, fancy wood, and what would be called today the stock's classic profile. Model 10s were also chambered for .303, .30–06 and .35 Winchester.

145-grain bullet 3,150 fps and a 180-grain 2,800. Note that these figures practically duplicate our modern 7mm Magnum. Though not so called, the .280 Ross was probably the first magnum.

Data associated with Ross are apt to be as variable as a chameleon's colors. I've seen the weight of his lighter bullet given as 139, 140, 142, 143, 145, and 146 grains! Velocities varied like-

wise. The cited 145-grainer at 3,150 fps is given in a 1912 Ross catalog for the famous Ross copper-tube bullet, manufactured, I believe, by the U.S. Cartridge Co. Apparently it first came out in 1908, years ahead of the .250–3000 Savage as the first commercial cartridge to reach or pass the magic milepost of 3,000 fps.

Astounding velocity for those times, and its

bullet performance was even more so. Other sporting cartridges of the early 1900s had top velocities of 2,000 to 2,200 fps. Their bullets were simple blunt-bowed types with large lead noses. True, the new .30–06 military load sent its 150-grain spitzer bullet off at 2,700 fps, but no decent expanding bullets existed for it. To make use of its high velocity and flat trajectory, hunters used army hard points which, when they hit game, darted and dived erratically. A bullet entering the shoulder might come out the flank, the back, or the foreleg. It might kill quickly or show no effect at all.

In his .280 copper tube, Ross produced a projectile that shot amazingly flat, dropping only 20 inches over 400 yards where other popular loads fell as much as 15 *feet*. It mushroomed well when it hit, and went in straight, making such terrific wounds that soft-skinned animals often collapsed as if struck by lightning. "Will shoot a bear's paw right off, or let daylight clear through him sideways," averred one enthusiast. (Reading this as a small boy, I thought it meant you'd see through him!) "Kills like dynamite," was a frequent verdict.

For grizzlies in the Rockies, stags in Scotland, antelope in Africa, mixed bag in India, the Ross became quite the rage. The staid British fell in love with it and fabled firms such as Lancaster, Holland & Holland, Jeffery, and later Rigby, all produced similar cartridges. Called .275s or .280s most were bored .276, with a groove diameter of .284, as are all 7mms. The Ross was a true .280, with a bullet diameter of .288.

Its popularity suffered a setback when Sir George Grey, scion of an old English family, tried to stop a charging lion with one in East Africa. Apparently the bullets either glanced off or broke up prematurely on the lion's back-sloping, hard-boned skull, and Grey was mauled to death. Possibly it would have been a different story had he used the long, heavy, 180-grain steel-jacketed bullet that was also available for the .280 at that time.

After that, there was heated debate about the reliability of the Ross on game. Modern high velocity calibers also seem to kill like lightning on some shots, fail on others, but the consensus of its users seems to have been that the .280 copper tube was one of the most effective bullets ever made for soft-skinned big game.

One Ross advertisement showed a hunter with three grizzlies killed in less than a minute. Another showed a wild sheep killed with an iron-sighted .280 at the range of one mile. This was supposed to demonstrate its extreme accuracy. I suspect it was supreme luck.

Undoubtedly the military match rifles Ross made in .280 caliber were, with special 180-grain target ammunition, very accurate indeed. At Brit-

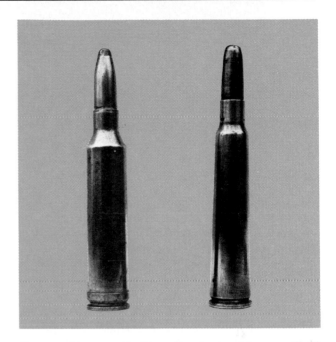

Compared here are the 7mm Remington Magnum (left) and .280 Ross cartridges. The Ross case, though longer, had a more pronounced taper and, consequently, it had less capacity.

ain's prestigious Bisley target matches they swept the boards in 1908 and won every first, second, and third prize in 1913, a record never equaled before or since. They may well have been the most accurate military-type rifles ever made.

What Ross sporting rifles would do is another question, and a matter of much dispute. That noted authority of the 1920s and '30s, Captain E.C. Crossman, claimed eight- and even six-inch groups at 500 yards with his .280 sporter. On the other hand, the revered authority Colonel Townsend Whelen wrote that the best he got with a Ross was 12-inch groups at 200 yards. Other Ross users were similarly divided. Only time I ever personally fired a .280 its bullets scattered all over the target, but the cartridges were old and I blamed them.

In many ways, those Ross sporters were lovely guns. Well-shaped stocks, beautiful finish, good balance, crisp trigger pull, and the smooth, swift, straight-pull bolt. Crossman called his "the rifle of my dreams." All Rosses had some neat extra touches: a roller bearing under the bolt, a safety lug which lifted just ahead of it to lock the bolt when the trigger was pulled, folding aperture sights on receiver bridges. A cute touch in early models, though not the .280, was the "Harrison loading ramp," a sort of button beside the barrel near the breech which, when pressed, lowered the magazine floor so that cartridges could be dumped in all at once. Another oddity in early

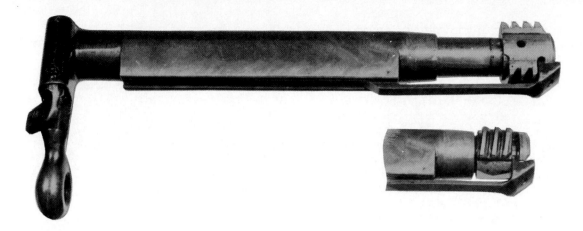

The heart of the problem: Model 1910 bolt, unlocked (top) and locked (right). When unlocked, turning the head of the bolt to the right would let it snap back into the bolt sleeve. So configured, the lugs wouldn't mate with the raceways and the bolt couldn't be reinserted into the action. If the head was rotated to the left, it would retract into the sleeve but because the lugs were horizontal, it could be shoved back into the action. Because the head was already retracted, the lugs couldn't extend into the receiver. Although the action was closed, and it could be fired, it wasn't locked. At least one death and several serious injuries resulted from that design flaw.

models, but not the .280, was that barrels did not screw into breeches but were fastened by two lugs.

Ross actions were very strong. Stamped on the .280 barrel was "Proved to 28 tons."—probably British long tons of 2,240 pounds each. In tests, Ross rifles are said to have withstood pressures of nearly 150,000 c.u.p.

But the action had serious flaws. Before going into them, a little background information is in order.

A striking personality was Sir Charles Henry Augustus Frederick Lockhard Ross, 1872–1942, Scottish laird, British baronet. Heir to great wealth, he spent it wildly in his youth. When he turned 21 he sued his mother for letting him waste it! At the English court he tutored a prince, consorted with actresses and ladies-in-waiting, contracted *that* disease, married, divorced, married again—twice. Owned a distillery in Scotland, mining interests in China and Africa, fought in the Boer War, fished in Labrador, hunted tigers—"bloody little beasts"—in India. In Canada he established the West Kootenay Power and Light Company and built its hydro dam that still supplies electricity to southeast British Columbians, including me.

An inventive genius fascinated by guns, he designed his first while still an Eton schoolboy. He was only 25 when he set up a factory in Hartford, Connecticut, to make a later design commercially. Not much came of it, but in 1901, when the Canadian government decided to break British apron strings and produce its own rifles for its armed forces, Ross submitted a design and won a contract. With $500,000 of his own money he built a factory in Quebec City and began to produce both sporting rifles and military models in .303 British.

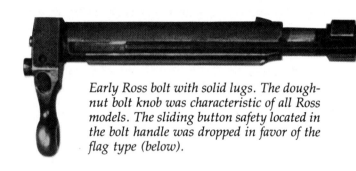

Early Ross bolt with solid lugs. The doughnut bolt knob was characteristic of all Ross models. The sliding button safety located in the bolt handle was dropped in favor of the flag type (below).

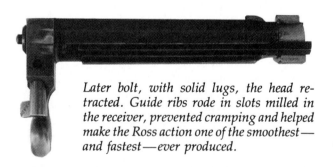

Later bolt, with solid lugs, the head retracted. Guide ribs rode in slots milled in the receiver, prevented cramping and helped make the Ross action one of the smoothest—and fastest—ever produced.

At first they were full of faults. The first thousand .303s went to the Royal North West Mounted Police, and blasphemy in that famous force broke all objurgatory records. Parts of the rifles broke, actions jammed, pieces fell off. When butts hit ground or floor hard in rifle drill, bolts fell out. At target practice in 1906, a bolt blew back into Sgt. Major Bowdridge's eye, blinding it. When Rosses reached the militia, one burst at the breech and killed its shooter. A critic in Parliament charged that the newly adopted Canadian rifle "kills as much behind as in front."

Full of new ideas and restless energy, Ross added many good details to his rifles, but impatiently ignored others. He did greatly improve those first Mark I rifles. His Mark II was a great advance and he kept adding touches, with a star for each one—Mark II*, Mark II**, and so on, until there were enough for a firmament. His Mark II**, introduced in 1907, seems to have been better than any before or after, and was in many ways an excellent rifle for a peacetime army, winning many shooting competitions. If only he hadn't kept changing!

By 1907 his chief interest had already shifted from the military .303 to the .280 cartridge. It was designed for him, and the Eley cartridge makers of England, by F.W. Jones, according to Barnes's *Cartridges of the World.* The same author says that in the same year, Eley began producing it in commercial quantities, soon joined by American firms. For this new caliber, Ross developed a slightly different action. It was straight pull, like all his rifles, but the barrel screwed into the breech and the bolt had serrated multiple locking lugs like those used today by Weatherby and other high quality rifles.

Later he incorporated the new action in a military .303, the Mark III, for the army. He felt it would stand higher pressures, and make the rifle adequate for the powerful .280 cartridge if he could succeed, as he hoped, in getting the Canadian and British armies to adopt it. Its high velocity and flat trajectory made it surer of scoring hits at long range than the .303 British at only 2,440 fps. Britain was planning to change to a .276 anyway.

The outbreak of World War I in 1914 wrecked Ross's dreams. Britain shelved all .276 or .280 ideas and went into the war with the tried and true Lee Enfield .303. The Canadians promptly joined them with their Rosses, mostly Mark IIIs, of the same caliber, and ran into trouble.

A turn-bolt action has powerful camming leverage for extraction; a straight pull has only the lesser twisting power of the helical grooves and ribs on inner bolt and outer sleeve. They work all right in today's Brownings and Remingtons, which have sharper helical curves than Rosses did. Apparently there were occasional Ross extraction troubles from the beginning. For example, when Colonel Whelen conducted his 100-round test of the .280 Ross, he reported six fired cartridges stuck badly and four jammed tight. This, despite that cartridge case boasting more taper than any other in use at the time. Other shooters had similar troubles, some none at all—a typical Ross disagreement.

It became very serious when the Canadian army took their Rosses into the trenches. What with the hastily made, wartime ammunition's soft or oversized cases, the ever-present Flanders mud getting into the complicated two-piece bolts and locking-lug seats, failures to extract became only too common. Soldiers were seen, with Ross butts on the ground in the trenches, stamping on the bolt handles to force them open, weeping with frustration and cursing Ross and his jam-prone rifles in the face of German attacks.

Ross had his factory bore larger rifle chambers until the empties which came out of them were swollen as if in advanced pregnancy, but they still jammed. And their 30.5-inch barrels and nine-plus pounds made the Rosses unwieldy in the cramped quarters of the trenches. In 1916 they were officially withdrawn and replaced by the handy, reliable British Lee Enfields. A vast sigh of relief probably soared up from the entire Canadian army. Only the snipers may have mourned the Ross for its accuracy.

His military contracts cancelled, Ross closed his factory. He didn't suffer—the Canadian government reimbursed him to the tune of $2,000,000—but he did cease all rifle production. Some of the leftover guns were scrapped; some were put into storage and reissued in World War II wherever there were shortages. After that war ended, some, perhaps all remaining, were put up for sale to the public at eight to ten dollars apiece. At that price there was a considerable revival of interest

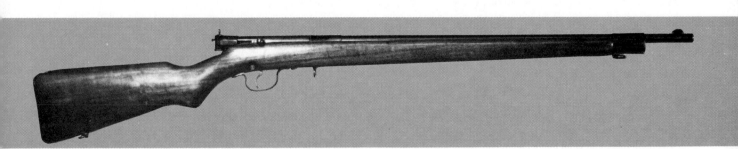

This is the single-shot .22 trainer, the Cadet. Like other Ross models, it featured a straight-pull bolt action. A sporting version was also produced.

Bolt designs of the four rifles below were inspired by the Ross concept of threaded locking lugs and a bolt within a bolt. Note the strong family resemblance between the BAR's bolthead (right) and that of the Ross Model 10 bolt shown back on page 98.

Browning BLR

Remington Model 742 Woodmaster

Remington Model 760 Gamemaster

Browning BAR

in Rosses, which fanned into flame an old and vexing question. Was the Ross safe to shoot? Or was it, besides being tops, or near it, in accuracy, killing power, velocity, and speed of action (the Russians won the Olympic rapid-fire match in 1954 with Rosses!) as dangerous as a rattlesnake?

Wherever old soldiers or old target shooters met, there were shouting arguments, and for some years they raged in the pages of the *American Rifleman.* Young and brash then, I foolishly horned in on it with an article passing on praises of the Ross sung by a veteran friend who had been an armorer sergeant in the Canadian army in both wars and handled Rosses by the hundreds. He called them safe as houses. Both of us were promptly slapped down by another ex-armorer-sergeant, Lindsay Elliott, who claimed to have known three soldiers killed by Rosses blowing back in their faces, and others injured. He cited a number of civilian cases, including the killing of Louis LaVallee in Alberta.

Elliott, in turn, was answered by many Ross defenders who, in years of experience with them, had never heard of any bolts blown out. They believed the accidents were the result of hang-fires, defective ammunition delaying firing until the bolt was partly pulled back and unlocked. They insisted the shooter's hand *must* have been gripping the bolt handle or it would have been injured.

It was generally admitted that, if a shooter was crazy enough to take his Ross bolt apart, it was possible to reassemble it so that it would go into the rifle and close without locking. In the army pins were put in to make this impossible. Well-

known Canadian gunsmith Ellwood Epps had done this for civilians, but considers that, when handled sensibly, a Ross bolt is entirely safe.

That Ross bolt is like a snapping turtle. When withdrawn from the rifle for cleaning or whatever, the bolt head sticks out of its sleeve nearly an inch, poised against the pressure of the mainspring. Turn the head to the right a mere touch, and snick! It snaps back into the sleeve, rotating so that it won't fit back into the rifle until pulled out and poised again.

In some Rosses you can turn the projecting bolt head to the left, too. Then it will snick back into the sleeve in such a position it *will* go back into the rifle. It can be closed, will not lock but can be fired. In short, all bolts don't have to be disassembled to be dangerous. Indeed, Lindsay Elliott claimed most .303 Mark III and .280 sporter Ross rifles could have their bolts turned into potential killers at the flick of a finger, although the Mark II Rosses with solid lugs and flush magazines seem safe enough.

A shooting pal of mind hunted with one of those handsome .280 sporters for over thirty years and had no fault with it. Probably there are others still in use, as are many .303 Rosses. Still, after all Lindsay Elliott told me, even though he may have exaggerated, I would not shoot a .280 again myself.

What a pity Sir Charles Ross did not stay with his design long enough to eliminate the jamming, and make it impossible to get the bolt in wrong. His rifle and great cartridge might have ushered in our Magnum Era fifty years before it came.

The Father of High Velocity

Col. Charles Askins

Ordinarily we are accustomed to thinking of Roy Weatherby as the father of high velocity. The West Coast impresario had a predecessor who may have given the inimitable Weatherby some ideas as to high-speed rounds. This fellow was named Newton. If we can award any accolades for high velocity, it rightfully goes to this remarkable individual.

Chas. Newton—he always signed his name as Chas. and never as Charles—was a gifted designer of not only rifles but even more spectacularly new cartridges for his rifles. Directly after the turn of the century his rifle stood pressures of 50,000 pounds per square inch, this in the days when it was thought that about 40,000 psi was absolutely the limit. His seven-lug interrupted-thread bolt was innovative, and his plunger-type ejector was many years later incorporated in most of our modern bolt-action rifles. And the best part of the story is that the Newton ejector, unlike those common today, only protruded through the bolt face as the bolt reached its final backward travel.

But quite as impressive as the Newton rifle were the Newton cartridges. These were in the .256, the .30, and the .35 Newton calibers. The .256 drove a 140-grain bullet at 3,000 fps; the .30 Newton forced the 172-grain ball to zip along at 3,000 fps; and the .35 Newton with 250-grain bullet made 2,975 fps. This was in 1915, now some 70 years in the limbo of the past. In those days the fastest round was the .280 Ross, which was good for 2,900 fps., but shooters were more accustomed to the .30–40 Krag and the .30–06 which rocked along at velocities safely in the 2,000-fps bracket.

The .30 Newton with its 172-grain bullet at 3,000 fps was truly sensational. This was 1915, remember, and the 06, then quite new and largely un-

tried, fired a 220-grain ball at 2,250 fps. The .30–30, popular choice of virtually every hunter west of the Mississippi, had a top speed of 1,950 fps. The .35 Newton with its 250-grain bullet frightened shooters in those days, just as all the .35 calibers do today. Energy amounted to 4,830 foot-pounds, a good rifle for Africa verily.

Bruce M. Jennings, Jr., of Sheridan, Wyoming, probably the greatest admirer of the gifted Chas. Newton, has written a book entitled *Charles Newton, Father of High Velocity.* This is the complete story, and while Jennings modestly states he has simply compiled the text from existing data on the firearms designer, in truth the story would be incomplete without the comment, additions, and observations provided by the man from Wyoming. There are 1,000 copies, and the cost is $40 per book. I am indebted to Jennings for his generosity in loaning me this tome.

An indication of the advanced design capabilities of Chas. Newton is that during the period 1909–12 he designed both the .25–06 and the .280

This article first appeared in American Rifleman

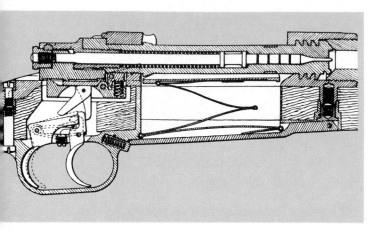

Newton's cartridges were matched with an equally innovative action with its interrupted-thread lock-up and novel takedown.

cartridges. Along with these sterling rounds came also the .22 Hi-Power and the .250–3000, the latter still with us as the .250 Savage. The .22 Hi-Power has pretty well fallen by the way. Savage took up both cartridges, and while Newton wanted to fire the 100-grain bullet in the .250 and expected to realize 2,800 fps, the Savage engineers dropped its weight to 87 grains and thus secured 3,000 fps. This was for the better publicity.

The .22 Hi-Power was achieved through a series of trial rounds. First Newton utilized the .25–25 Stevens case, then he went to the .28–30, and finally settled on the .25–35 Winchester which was then standardized. The 70-grain factory load rocks along at 2,800 fps.

During that intensive period, 1906–12, Newton, besides working up the .25–06 and the .280, also developed three different cartridges, all in .30 caliber. One of these was the .30 Newton Express made up on the necked-down .40–90 case. This round was intended for single-shot rifles and drillings. Another .30 caliber was the idea of Fred Adolph of Genoa, New York, a custom maker of very fine double rifles and drillings. He persuaded Newton to neck down and reform the 11.2mm Schuler. This casing has a rebated head and the caliber is about a .44. It is an elephant killer in its original form. Newton necked it down to .30 caliber. The .404 Jeffery case was shortened and reshaped and was then sized down to take the .30 caliber bullet. Both these big cartridges were intended for bolt-action rifles. The latter cartridge, made up on the .404 Jeffery, was dubbed the .30 Adolph Express.

After this developmental work, Newton formed a corporation in August 1914, at Buffalo, New York, his hometown. Interestingly, Chas. Newton was a lawyer, not an engineer.

During this same year our firearms designer negotiated with the Mauserwerke at Oberndorf for 24 rifles to be delivered by August 15. These rifles were to be chambered for the .30 caliber, the 8mm and the .35 Adolph Express. A second shipment of 24 rifles was scheduled for September 15. This following order was to include the .256, the .30, and the .35 Newton. It is believed the first 24 rifles were probably, at least most of them, for the Adolph cartridges, the 11.2mm and the .404 as altered.

The .308 Norma Magnum with 180-grain bullet gallops along at 3,100 fps in the present-day loading. Newton did this three-quarters of a century ago. The .358 Norma acknowledges 2,890 fps with the 250-grain bullet The .35 Newton was listed at 2,975 fps with 250-grain bullet. To me it is more impressive that this lawyer-experimenter-manufacturer was able to accomplish these exemplary advancements in cartridge design so early.

The true Newton cartridges, Jennings believes, were designed by Newton, and he turned to the U.S. Cartridge Co., Lowell, Massachusetts, for their manufacture. These rounds had a rim diameter of .525 inch, while the Schuler went .470 inch and the Jeffery a full .555 inch.

Germany cranked up World War I during mid-August 1914, and it is believed only one shipment of the 24 rifles was ever received.

Newton then went to an American manufacturer, believed to be Marlin, and contracted for a quantity of barrels and stocks in the .256 caliber. It was his intention to use the 03 Springfield receiver. Stocks and barrels were listed in the Newton Arms Co. catalog at $12.50 for the tubes and a like sum for the stocks. My father, Major Charles Askins, had one of these rifles. He attached a Lyman 48 receiver sight, and this was the only way the rifle could be told from the as-issue military weapon.

The Newton catalog of 1915 stated that the .280 caliber would be similar to the .280 Ross except that the casing would be somewhat shorter and fatter. This catalog also announced for the first time the .33 Newton. It was intended to drive a 200-grain bullet at 3,000 fps. As it finally transpired the company never produced the .280 or the .33 calibers.

While it appears U.S. Cartridge Co. loaded the first Newton rounds, probably experimental loadings, in 1915 the entrepreneur mentioned in his publication that the Union Metallic Cartridge Co. (now Remington) was turning out both his cases and his bullet jackets. The Newton Co. loaded the cartridges themselves. The Newton cartridges were with either DuPont No. 10 or No. 15 propellants. In the .30 caliber, a charge of 65 grains was standarized; in the .35 Newton the loading was 67 grains.

By this time the Buffalo plant was producing 125 rifles weekly. I am indebted to Frank de Haas, author of *Bolt Action Rifles*, for a description of the Newton firearm. Frank, who was a Contributing Editor to the *American Rifleman* from 1959 to 1975, has this to say, paraphrasing: The action was a turn-bolt design with staggered column box magazine. The action featured a one-piece bolt with interrupted locking lugs at the front end. The bolt handle and safety needed no alterations for the mounting of a low scope sight. There were double set triggers and a hinged floorplate together with a takedown system. The extractor was the conventional Mauser kind. The ejector was a spring-actuated blade which has been copied pretty widely in more recent years.

The receiver shows considerable resemblance to the Mauser 98. It is a one-piece forging and is flat-bottomed except for a short length at the front end of the receiver ring. The bolt is a one-piece, undoubtedly is a forging, and is of machined construction. At the front end of the bolt is a set of dual locking lugs, four narrow lugs on the left side and three lugs on the right. The four lugs lock into the top of the receiver ring and the three lugs lock into the bottom of the receiver. Each set of lugs projects beyond the bolt body and this requires a raceway be made in the receiver for the bolt to move rearward and forward. The extractor is of the Mauser type which does not rotate as the bolt is opened and closed. It is held in place by a collar. A lip just to the rear of the extractor hook engages in a groove in the ledge of the bolt-head. This is the reason there are only three lugs on the right side of the bolt. This to make way for this lip, DeHaas notes.

The bolt has the usual Mauser recessed bolt face with part of the rim cut away. As the bolt feeds a round up out of the magazine the head of the cartridge rises directly into the recess and becomes engaged beneath the extractor hook at that time. This feature prevents double loading which means the bolt cannot feed another cartridge from the magazine until it has gotten rid of the first round. This is a most worthwhile feature of a hunting rifle.

There are two small safety lugs; these turn down just ahead of the receiver bridge. The Newton safety and striker are lighter, faster, neater, and handier than either the 98 Mauser or the 03 Springfield, DeHaas writes. The bolt stop and ejector are mounted in a cross pin in the bottom center of the receiver bridge. Both are provided tension by a coil spring. The bolt is stopped during its rearward movement when the stop moves into a deep notch near the front of the bolt. At the same time the ejector part of the bolt stop slides upward into the narrow slot cut in the bolt face and thus ejects the empty.

The bolt stop is linked to the front lever in the sear so that pulling the front trigger (this is a double-trigger rifle, remember) as far as it will go swings the bolt stop down so that the bolt can be removed from the rifle.

There is a hinged floorplate which is released and drops downward by pressure on a button located on the front outside surface of the trigger-guard. The front guard screw passes through the magazine opening and screws into the recoil lug of the receiver. The receiver tang is fitted so that the barreled action may be lifted out of the stock, DeHaas observes.

Newton rifles, regardless of caliber, weighed from 7½ to 7¾ pounds. The standard barrel length was 24 inches, although Jennings has a Newton

Newton case and rim types abound. These headstamps are .30 Adolph (Express), pair of .30 Newtons, and .35 Newton.

Original Newton "heat-insulated" bullets included (left to right) 80- and 100-grain two-diameter .22s, a 129-grain .256, a pair of 172-grain .30s, and two of the 250-grain 35s. Sectioned bullets exhibit Newton's wire reinforcement.

This engraved .256 caliber Newton prototype rifle resides in the NRA Firearms Museum. The rifle was a gift of Newton's friend A. D. Bissel.

with 26-inch barrel and another with 30-inch tube. The stocks had 14-inch length of pull, and there was an aperture sight which could be attached to the cocking piece. Sling swivels and a sling were available. The rifle sold for $40, but there were two fancier grades which boosted the price slightly.

The United States went to war April 6, 1917, and the Newton supply of empty cases and bullet jackets ceased as all America's resources were devoted to the war effort. The company by this time was producing a reputed 500 rifles monthly in its Buffalo facility. Undaunted by the shortages of cartridge cases, Newton scurried around and by the middle of April 1918, a year later, the firm had its own machines for the fabrication of the inimitable Newton round. Then the company went into bankruptcy.

The rifle manufacture closed down during August; the cartridge plant kept on until the early days of 1919. The company in the hands of the receiver finished about 1,200 rifles, with an additional 400 ready for assembly when it in turn folded up. The original Newton Co. had manufactured about 2,400 rifles, with an additional 400 ready for assembly when it in turn folded up. The original Newton Co. had manufactured about 2,400 rifles before it went into receivership. So altogether there were about 4,000 Newton rifles in existence. There were about 1,000 rifles that were rejects. The new Newton Arms Co. was located in Brooklyn with sales offices in the Woolworth Building, New York City. The new owners proposed to sell the 1,000 rejected rifles for $76 for the .256 and .30 and $81 for the .35 Newton. There was also ammunition for sale. At the time the original company went under, Newton stated that he had a production capacity of 40 rifles per nine-hour day and 15,000 cartridges.

It is interesting to realize that beginning in the early 1920s, the Western Cartridge Co. regularly loaded .256, the .30, and the .35 Newton. The .280 was never manufactured.

The bankruptcy receivers proposed to use the Newton Arms Co. name in the further manufacture of the Newton rifle, most especially in the sale of the 1,000 condemned rifles. Chas. Newton brought suit against the receiver objecting to the use of his name. He won the lawsuit. Directly after that the new firm went into bankruptcy, and at that time it still had 250 of the rejected rifles on hand. These were taken over by Kirtland Bros. of New York City and were all sold.

In April 1919, Newton organized his second company. It was dubbed the Chas. Newton Rifle Corp. The plan was to manufacture the original rifle with about 20 improvements. There was a new set trigger, a bolt stop as on the later model Newton Buffalo rifle, safety on the left side of the bolt sleeve, parabolic rifling, plunger-type ejector collar eliminated, and other changes. There were plans to buy tools from the Eddystone Arsenal, but this came to naught.

The new rifle was on the Mauser action with the Mannlicher-style bolt handle, new-style reversed double set triggers, three-leaf express sights, the front sight milled as an integral part of the barrel, and a newly designed Newton stock. Some of these rifles had a cloverleaf milled into the barrel muzzle. This was supposed to eliminate tipping of the bullet. The rifling was parabolic in some rifles and the more conventional rifling in others. These rifles were made in Germany and went on sale in 1922. The rifle was known as the Model 1922. It is believed by Bruce Jennings that probably Newton got about 100 of these rifles from Mauser. The second company went out of business finally because of a lack of working capital.

Charles Newton, Father of High Velocity, is available from the author, Bruce M. Jennings, Jr., 70 Metz Road, Sheridan, WY 82801, at $40. *Bolt Action Rifles* (revised edition), by Frank de Haas, is $14.95, postpaid, from DBI Books, 1 Northfield Plaza, Northfield, IL 60093.

A Man Named Weatherby

Jim Carmichel

The most influential personality in sporting arms and ammunition design during this half of the twentieth century is Roy Weatherby. It's as simple as that. There really isn't a second place because Weatherby's contributions to modern ballistics and rifle design are both unique. Had he achieved one without the other, he would still be a towering figure in the industry. Everyone who designs, makes, sells, buys, and uses sporting rifles and ammunition is somehow touched by Roy Weatherby. It doesn't matter whether or not we like the big magnum concept or the "Weatherby look," we all react to Weatherby in some way. Even manufacturers who make conservative or "classic" rifles are reacting to Weatherby, which is just another way of being touched by his influence.

For a man who has succeeded so well, whose ideas have influenced so many, whose bullets split the air like lightning, and whose guns sizzle the senses like abstract art. Roy Weatherby is not what you'd expect him to be. He's a super gun nut who doesn't talk like a gun nut. Rather than talking about high velocity and knock-down power, you quickly find yourself talking to Roy about your home and family, and he might even show you some pictures of his family. Pretty soon, he writes

This article first appeared in Outdoor Life

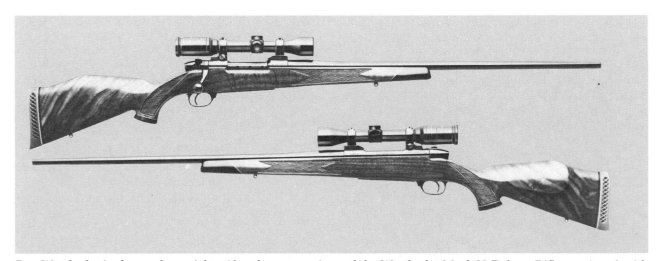

Roy Weatherby is shown above right. Also shown are views of the Weatherby Mark V Deluxe Rifle, equipped with 3–9 × 44 Weatherby Supreme scope on a Buehler mount.

Personally testing his products over the years, Roy Weatherby poses with a black rhino he took in the 1950s.

your name in his notebook because he sincerely doesn't want to forget it. If the discussion does turn to guns and ballistics, you'll do most of the talking and he'll listen with keen interest, probably making a note or two on odd scraps of paper. Somehow, he gives you the impression that you know more about the topic than he does, even when you're talking about his products. Days later, after you've forgotten the conversation and you're sure that he has, you may get a personal letter on the distinctive Weatherby stationery, analyzing in detail the main points of the conversation. Yes, he's a gun nut but, more than that, he's a people nut. That's why he's one of the most liked and respected men in the business. Kings, princes, the rich and the famous may beat a path to his door but, when he reaches out, it's to shooters and hunters. Guys like himself—you and me.

Roy was a handloader and wildcatter before most of today's generation of handloaders were born and, like other wildcatters before and since, his primary objective was increased bullet speed. In the time-honored manner, he achieved this by the simple expedient of burning a lot of powder in a big case behind a relatively small bullet. Why, then, did Weatherby's ideas about high velocity succeed on a worldwide scale while similar cartridges were tried and forgotten. The key is Weatherby himself. He's a salesman who believes in a good idea and a good product. In 1945, he believed that he had a good idea and he was ready

to put it to the test. Well, that was 40 years ago and what happened, as they say, is history—shooting history.

THE EARLY DAYS

Weatherby started out on a farm near Salina, Kansas, and, after finishing school, worked for Bell Telephone. He must have been good at what he did because, by 1937, he had worked his way up to a managerial position. He had bigger dreams, however, and went to California, the land of dreams, where he became an insurance salesman. Again he succeeded in a pretty big way, earning a depression-era salary of $700 per month. That was big dough in those days, and tells us that Weatherby had become one hell of a salesman, a skill that was to serve him well in later years.

By the late 1930s, Weatherby was more than what you would call a serious handloader. He was already experimenting with high-velocity cartridges and redesigning existing calibers. One of his first ventures into the exotic world of ultravelocity was the .220 Weatherby Rocket. This wildcat, developed in 1943, was never made available as factory ammunition but was chambered in custom rifles. It was made by increasing the powder capacity of the .220 Swift case.

By then, Weatherby's ideas about high velocity and cartridge-case design were coming in rapid-fire order. Building on Ralph W. Miller's "venturi" case-neck shape, Weatherby added the rounded shoulder shape that has become the hallmark of his magnum case design. The idea of Miller's "venturi" neck design with a radiused juncture at the neck was to facilitate gas flow from the case into the barrel. Weatherby reasoned that, if a curved corner at the neck would speed up the gas flow, it would also be a good idea to help the powder gases "turn the corner" at a round shoulder.

Whether or not the double-radiused shoulder benefits performance is the sort of thing ballistic buffs love to debate but, from the standpoint of promotion and sales, it was a stroke of shining genius. The distinctive shape makes Weatherby's cartridges the most recognizable in the world. Weatherby's other stroke of genius was his adaptation of the belted magnum case. He did not invent the belted magnum case, as some fans believe, nor did he coin the term "magnum." The belted case, so called because of the beltlike structure around the case head, was introduced back in 1912 by the British firm of Holland & Holland when they brought out their .375 Magnum cartridge. Though the belt looks like it reinforces the case so that it can withstand high pressure, the simple truth is that it contributes little, if anything, in the way of additional strength. It functions primarily as a means of headspacing. In any event, a belt has a high-tech look about it that appeals to shooters, and the word "Magnum" sings a song of its own. Put "Weatherby" and "Magnum" together and you have an anthem. Can you imagine a Smith Magnum or a Jones Magnum?

During those virile days of high-velocity experimentation, nearly all wildcat cartridges were based on the easily available .30-06 case. Weatherby's use of the basic H&H Magnum case allowed even more powder to be burned than could be stuffed into the slimmer '06 case, so his velocities were higher. The first definable Weatherby Magnum, for the record, was the .270 developed in 1942, followed in close order by the .257, 7mm, .300, and .375 Weatherby magnums. Each of these was based on the existing H&H Magnum case, used either full length or somewhat shortened.

RIFLE MAKING

Roy's next step was to produce Weatherby rifles capable of firing his razzle-dazzle cartridges. This was a big step because it meant jumping into the gun business with both feet, opening a shop, hiring workers, and designing a Weatherby rifle. It was 1945, hundreds of thousands of G.I.s were returning home and, again, Weatherby had an idea to work with.

When Weatherby opened his shop and went into the rifle-building business in 1945, one of his most pressing problems was the shortage of suitable actions. Like most custom shops before and since, he was forced to use whatever actions were available. He says that he preferred Winchester's Model 70 mechanism but that, in order to get one, he had to buy the whole rifle, thus running up his expenses. Hence, surplus Mauser, Enfield, and Springfield actions were routinely used in Weatherby rifles, plus whatever commercial actions found their way to his shop. Late in the 1950s, Weatherby began using Mauser-type actions newly manufactured in Belgium by FN and marked with the Weatherby name. By using the FN actions, Weatherby's rifles gained a much needed uniformity of form and manufacture and, by then, he was also manufacturing his own barrels. Still, though, there was difficulty about making Weatherby Magnum ammunition.

For several years after opening his shop, the production of Weatherby Magnum ammunition was pretty much a do-it-yourself proposition. The customer either loaded his own ammo or got someone to do it for him. Even the ammunition loaded and sold by Weatherby's company was

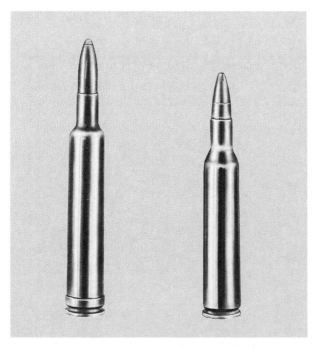

Here is a comparison of the rounded shoulder and belted head of the case on a .240 Weatherby Magnum (left) and the case of a 6mm Remington.

little more than a handloading operation. The new cases were actually fireformed in Weatherby chambers.

Realizing that, if his company was to play in the big league, he would have to have proper factory-loaded and easily available ammunition. Roy set about finding a manufacturer. American ammo makers scoffed at the idea of making ammo for the California upstart, so he took his ideas to the Swedish firm of Norma. What he told the Swedes is lost to history, but they must have been impressed. He had little credit and less capital but, working on faith and a handshake, Norma took on the job and produced the first legitimate Weatherby Magnum cartridges in 1953. Since then, they have manufactured virtually all Weatherby ammunition, and the relation has been a happy one. Several years ago, when I was visiting the Norma plant, the management had a lot of good things to say about Weatherby, the best being that he was an honest and reliable gentleman.

Like most successful businessmen, Roy has a knack for hiring the right people. When he hired ace stockmaker Leonard Mews to make the stocks for his rifles, the chemistry between the two men sparked. Mews, who was one of the best stockmakers in the country, had a flair for sleek, racy stock designs that meshed with what Weatherby had in mind. It's worth noting that some of Mews's original designs were so radical that Roy had to tone them down. The high Monte Carlo comb, so characteristic of Weatherby rifles, for example, was even higher, as made by Mews, until Roy lowered it somewhat. Other Weatherby features, such as the angled fore-end tip, tri-shaped forearm cross section, and flared pistol grip were designed by Weatherby himself. The trademark diamond grip cap inlay and, yes, let the record show, the white spacers that set off the fore-end tip and the grip cap were also the products of his imagination. Back in 1950, when the Weatherby stock had evolved into pretty much what it is today, the major rifle makers, here and abroad,

had themselves a giggle at Weatherby's radical stock design and ornamentation. They would have giggled a lot less could they have known the Weatherby stock was to become the most copied design of the century. If you doubt this, just get out a few gun catalogs and count all the guns, from rimfires on up, that copy one or more of Weatherby's features.

Back when I was a youngster, I bought every shooting book and magazine that I could lay hands on, and nothing made my heart beat faster than the photos of the lavishly carved, checkered, and inlaid Weatherby rifles. When I finally saw a Weatherby rifle in the flesh, I nearly swooned and, naturally, when I began making my own stocks they were, as nearly as I could make them, copies of Weatherbys.

THE .460 IS BORN

The only Weatherby cartridge to become obsolete was the .375 Magnum, and that was by intent. Realizing that the .375 caliber had greater potential than was permitted by his adaptation of the old H&H case, Weatherby designed a wholly new case with significantly greater powder capacity. The new round was christened the .378 Weatherby Magnum, and it was a sure 'nuff bone crusher. The old .375 H&H had a muzzle velocity of 2,740 fps with a 270-grain bullet, and the .375 Weatherby produced 2,940 fps, but the new .378 churned up no less than 3,180 fps for a muzzle energy of more than *three tons*!

That was in 1953, and Weatherby figured that he had designed the ultimate cartridge for hunting dangerous game. When he learned that Kenya was requiring that dangerous-game rifles be .400 caliber or larger. Weatherby opened up the neck of his huge .378 to .45 caliber. Thus, in 1953, was born the awesome .460 Weatherby Magnum. Launching a 500-grain bullet at 2,700 fps, the muzzle energy is 8,095 foot-pounds,

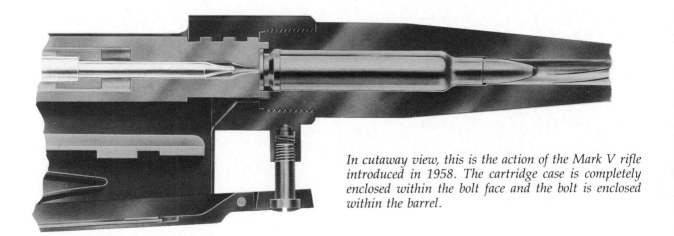

In cutaway view, this is the action of the Mark V rifle introduced in 1958. The cartridge case is completely enclosed within the bolt face and the bolt is enclosed within the barrel.

Above is a .240 Weatherby Magnum Mark V custom rifle with engraved barrel, receiver, scope mount, floor plate and triggerguard assembly. The myrtlewood stock features running-gazelle and forearm inlays. Another rifle shows the detail of an engraved receiver, barrel, and scope mount, as well as that of Weatherby custom stock work.

outstripping even the .600 Nitro Express (7,600 foot-pounds), which makes it the world's most powerful factory-loaded sporting cartridge.

The .378 and .460 cartridges were so big that they could only be handled by a special action, thus making even worse a state of affairs that had long been a burr under Weatherby's saddle. What he wanted was a distinctive action of his own design that would complement his cartridges. This became a reality in 1958 with the introduction of the Mark V action, a truly innovative mechanical design that redirected the future of turn-bolt development. Weatherby, working with his engineer, Fred Jennie, was deeply involved with the development of the Mark V action from the beginning, and he is personally responsible for many of its features. The Mark V has too many features to list here, not the least being the nine-lug locking arrangement (rather than the usual two). The outcome was neatly summed up several years ago when another rifle maker conducted a series of tests to determine which of today's actions has the most strength. The action that held together longest was the Weatherby Mark V.

Products that perform like Weatherby rifles and cartridges are bound to be controversial and, if there is controversy, there will be critics. Yet it is hard to criticize a Weatherby rifle or a Weatherby Magnum cartridge and make it stick. Anyone who

says that the rifles and ammo aren't accurate should try the fiberglass-stocked Fibermark rifle in .257 Weatherby Magnum. I used one on a pronghorn hunt a couple of years ago. With factory loads, it grouped five shots under an inch at 100 yards. The .240 and .224 magnums will do as well, and so will the big Weatherby Magnums, if you have the determination to hang on while they hammer your shoulder.

And when I next hunt elephants in the equatorial rain forests, I'll be carrying a .460 Weatherby Magnum. Shooting distances are only a few feet at most, and the .460 will put a big *tembo* down even if you can't get a brain shot. A professional big-game hunter I know uses the .460 because it will "shoot through an elephant lengthwise and totally unwind his mainspring," as he puts it.

A while back, when I was chatting with Roy, I asked if he considered himself a cartridge designer, a rifle designer, or a salesman.

"I'm just an idea man," he answered, "who knows the market, knows the product, and knows what people want to buy. My aim was to have the best and safest gun that anyone ever built, different from what they already had."

My guess is that, a century from now, when hunters and shooters talk about what went on in the twentieth century, the man they'll talk about most is Roy Weatherby.

PART FIVE

AMMUNITION, BALLISTICS, AND HANDLOADING

Of Prime Importance

Lon W. Sorensen

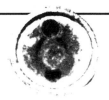
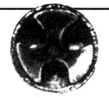
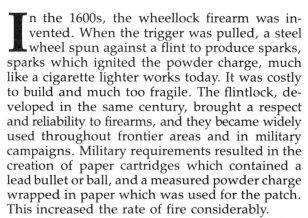

Here are three different types of anvils that were used in the Boxer primers.

In the 1600s, the wheellock firearm was invented. When the trigger was pulled, a steel wheel spun against a flint to produce sparks, sparks which ignited the powder charge, much like a cigarette lighter works today. It was costly to build and much too fragile. The flintlock, developed in the same century, brought a respect and reliability to firearms, and they became widely used throughout frontier areas and in military campaigns. Military requirements resulted in the creation of paper cartridges which contained a lead bullet or ball, and a measured powder charge wrapped in paper which was used for the patch. This increased the rate of fire considerably.

In 1805, Reverend Alexander Forsyth discovered that by mixing nitric acid, mercury, and alcohol, a heavy gray powder was the eventual result. The compound, called fulminate of mercury, was extremely sensitive to shock of any kind. When struck, it ignited with a flash. Another Englishmen, Joshua Shaw, developed a percussion cap system for guns based on the fulminate of mercury compound.

As the percussion system developed, it became more reliable than the flintlock, and was the beginning of the ignition system we know and use today. By 1840, the percussion rifle was used extensively and flintlocks were no longer being produced.

In the late 1850s, metallic cartridges began to appear in breechloading weapons. The fulminate of mercury compound was still in use, but there were serious questions about the best way to prime the new self-contained cartridges. Although many ideas were put forth, only two methods came to the front of the race in the 1860s.

Colonel Hiram Berdan, a noted Civil War sharpshooter and member of the U.S. Army Ordnance Corps, developed a primer using an anvil

This article first appeared in Handloader

Many handloaders prefer to use a priming tool built especially for primer seating. Here the primer is being inserted into the tool. (John Sill photo)

Here are two views of the Berdan primer pocket (left) and the Boxer primer pocket (right). Note the difference in flash-hole diameters.

BERDAN PRIMER   **BOXER PRIMER**

Anvil

Priming mixture   Priming mixture

Primer cup

that was an integral part of the cartridge case. There were two flash holes in the primer pocket, and a small copper cup held the priming mixture.

Edward M. Boxer, a British army officer, devised a system using a cartridge case with one large flash hole. Each primer was a small copper cup lined with priming mixture and containing its own anvil.

The Berdan system was adopted in Europe, largely because it could produce ammunition less expensively. After a great deal of experimentation, the U.S. Army chose the Boxer priming system. Reloaders throughout this country still applaud that decision because Boxer-primed cases are much easier to deprime. Berdan cases can be reloaded but it requires a special tool to deprime them.

Shortly after the priming systems were established, a major problem surfaced. After firing, the fulminate of mercury residue attacked the brass cases, breaking them down in a very short time. At long last, fulminate of mercury was relegated

to blasting use and centerfire primers were made with potassium chlorate for the priming compound. That solved the brass problem.

In the 1870s and '80s, smokeless powder was being introduced and subjected to endless experimentation. It was noticeably more difficult to ignite than black powder. By increasing the amount of chemical compound in the primers, the smokeless powder could be ignited satisfactorily but the more powerful nonmercuric primers presented a serious problem. After firing, the potassium chlorate left a residue in the barrel that was very similar to table salt and highly corrosive. Shooters who left their guns uncleaned for any length of time found the bores of their weapons rusted.

Civilian and military shooters learned to live with the problem, the military making a sustained effort to have firearms kept spotlessly clean. A clean, shiny bore was the first thing a buyer looked for when purchasing a used gun, a practice which makes good sense even today.

In the 1940s, commercial primer manufacturers began producing nonmercuric, noncorrosive primers using lead styphnate as the priming compound. It has proved to be exceptionally reliable and consistent, and greatly contributes to the

Shotgun primer components: battery cup (left), anvil (center), and primer cup.

SHOTSHELL PRIMERS

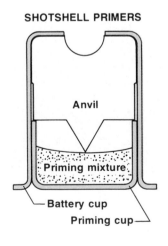

Anvil

Priming mixture

Battery cup

Priming cup

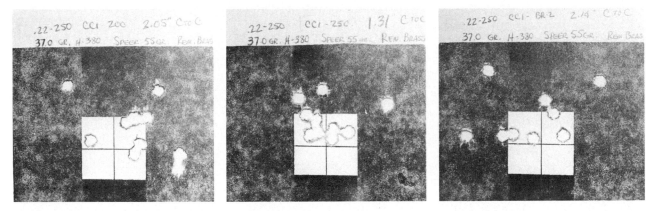

These three 10-shot groups were fired from a .22–250. Bullets and powder charges remained the same in all loads—only the primers were changed. Any questions?

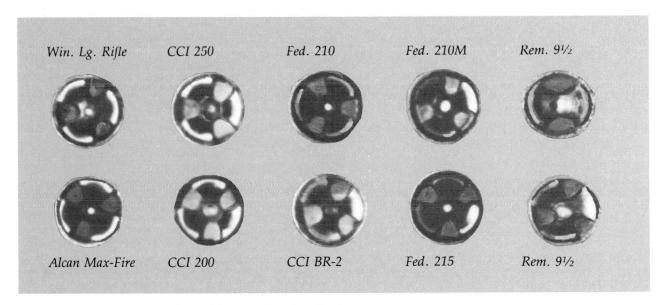

Selection of primers commonly used by American handloaders.

quality of the ammunition we use today.

Not all primers made since the 1940s are noncorrosive, unfortunately. The U.S. government didn't make the switch to nonmercuric, noncorrosive priming until the early 1950s. Except for Frankford Arsenal match ammunition, any government ammunition made after 1953 can be considered noncorrosive. Frankford Arsenal match ammunition continued to use corrosive primers until 1956. If you happen to inherit a batch of issue ammunition that has been around for 30 years or so, make sure your rifle gets a good, thorough cleaning soon after shooting it.

Now that primers are nonmercuric to preserve cartridge cases, and noncorrosive to save barrels, it is almost too easy to take the priming process for granted. Although the primer is the smallest and sometimes the least expensive part of the

reload, its proper function is essential to good performance for the reloader.

Consistency is the key to proper primer seating. A primer should be seated with uniform, firm pressure to push the anvil firmly against the bottom of the primer pocket. The primer cup's side wall can, but need not be, against the bottom of the primer pocket. The primer should be seated firmly enough to put a small amount of tension on the primer's explosive pellet caught between cup and anvil. If the anvil is not against the bottom of the primer pocket, the primer may move forward when struck by the firing pin, absorbing some of the firing pin's energy and causing inconsistent ignition. You can't tell you're shooting cartridges with improperly seated primers unless you get a hangfire or a misfire, but it will sure show in your targets.

When handling primers, hands should be clean, dry and—most important—free of oil. (John Sill color photos)

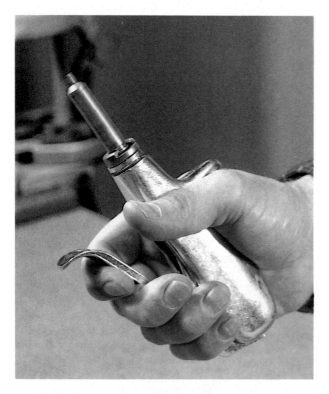

Left: Shown in use is the RCBS Posi-Prime hand-priming tool, seating small rifle benchrest primer in .22 CHeetah case. Right: Best primer seating depth is slightly below flush—.003- to .005-inch below rim. However, measuring depth precisely is difficult and unnecessary. To check for proper depth, merely run your finger over the case head to confirm that the primer is just below flush.

Protruding primers, in extreme cases, can cause feeding problems or difficulty in closing the action of lever, automatic, or pump rifles. Bolt-actions will be harder to close on protruding primers, sometimes feeling as if the bullets are seated too far out. In many instances this problem never develops because most presses and priming tools will not allow a case to be removed from the tool if the primer protrudes much at all.

Primers should be seated flush or slightly below the rim of the case. I've read that they should be .003 to .005 inch below the rim but measuring how deep they are seated is too much trouble. Run your finger over the bottom of each case after the primers are seated and you can tell if they are flush with the bottom of the case or below it.

Most reloading presses with primer arms have a little more leverage than many of the priming tools available and must be used with care so that primers are not crushed when seated. If the primer cup is crushed, the explosive pellet can crumble away from the critical area for proper ignition. If you do seat a primer with enough force to crush it, you may ignite the primer in the process. That is why instructions about proper priming encourage you to wear safety glasses.

Not all crushed primers will misfire, and not all misfires are caused by crushed primers. The number one reason for misfires is contamination of the priming mixture. Primers can be contaminated by moisture, body oils—even sweaty hands. Most common is resizing lubricant on the reloader's hands when he picks up a primer and

puts it in the cup of the priming arm of his press.

I have heard of misfires caused by an anvil falling out of a primer before being seated but have never experienced such a misfortune. Anvils come out easily after firing, but any that jar loose before firing would have to have been subjected to mighty rough handling. The reloader

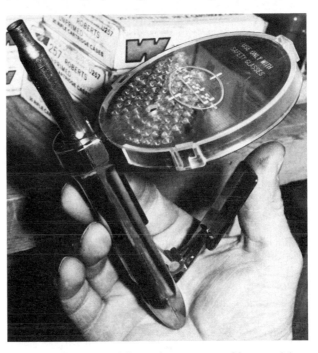

Lee Auto-Prime tool in action—a portable, sensitive, and inexpensive priming arrangement.

Below left: A most common method of priming a case. It works fine as long as fingers are clean and dry.
Below right: Pacific's automatic system with auto-feed priming—a no-touch approach.

 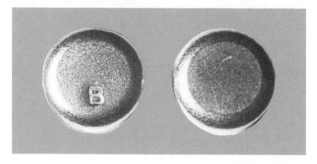

Left: When primer pockets are stretched from too many loads or distorted from improper reaming, they may leak. Note the smoked area between the primer cup and the pocket. Right: CCI Benchrest primers can be identified by the small B stamped on them.

can quickly check that anvils are present in each primer as he primes his cases.

There are several priming systems on the market today that will allow you to prime cases and never have to touch the primers. Bonanza, Lee, RCBS, and others offer priming tools that only prime cases. They won't resize, chamfer, seat bullets or anything else. They just seat primers and they do it very well.

The Bonanza and RCBS priming tools are bench models; the Lee Improved Priming Tool and Auto-Prime are both hand tools and are very portable. Many reloaders are not only impressed with the Auto-Prime's performance, but also its price. Priming tools are a good investment for the more consistent, precise reloader.

Primers leave some burned residue in the primer pocket when fired. With Boxer-primed cases, the residue is cleared from the flash hole by the primer pin when the case is deprimed. The primer pocket will have some carbonlike residue at the bottom of the pocket which is easily removed with a primer pocket cleaning tool. If the pocket is not cleaned, a problem seldom occurs but in extreme cases, it can prevent the new primer from being seated flush or below the rim of the case. For reloaders who pride themselves in producing ammunition as good or better than factory loads, it only takes a moment to clean a primer pocket.

Primer pocket reamers have been used as primer pocket cleaning tools by some reloaders. Reamers are used to remove the crimp around the primer pocket of government brass, but GI cases need to be reamed only the first time they are reloaded. Too much reaming, or reaming the pocket crookedly, will enlarge the pockets and may result in blown primers or allow gas to leak from the primer pocket when firing. As experienced reloaders know, cases with enlarged primer pockets should be discarded. Primer pocket reamers have been replaced in some manufacturer's product lines with primer pocket swagers, which use the reloading press's leverage to swage too-small pock-

ets to proper size. They are virtually foolproof. RCBS and Bonanza both make primer pocket swaging tools, and there may be others.

Large Rifle and Large Pistol primers are both made in .210-inch diameter, while Small Rifle and Small Pistol primers have .175-inch diameters. Pistol primers are designed to ignite small charges of fast burning powder, and to do so with a relatively light blow to the primer. Rifle primers must ignite large charges of slow burning powder but have to withstand a fairly heavy firing pin blow. Consequently, rifle primers are made with heavier brass cups and a more explosive compound than pistol primers.

If pistol primers are used in rifles, the heavier blow of the firing pin sometimes perforates the primer and allows gas to leak back into the action, and into the shooter's face.

If rifle primers are used in pistols, misfires or inconsistent ignition may result from the lighter firing pin blows. Rifle primers can be used in centerfire pistols designed to operate at high pressures like Thompson/Center's Contender, Remington's XP-100, and Wichita's Unlimited Silhouette Pistol.

Magnum primers, designed to ignite heavy charges of slow burning powders, contain more explosive compound than standard primers. That gives more consistent combustion in magnum cases, and can be very beneficial when using Ball powders, which are a bit harder to ignite. I've had excellent results using magnum primers with Ball powders in cases as small as the .22–250. Magnum primers are especially helpful in hunting loads that are to be used in temperatures below freezing, giving the hunter the best possible cold weather performance.

Benchrest primers are not made of any different materials than regular primers; instead they have a higher degree of uniformity and consistency. The extra consistency can be a real benefit for those who weigh each charge of powder, each brass case, and each bullet to obtain sub-MOA

PRIMER DESIGNATIONS

	Large Rifle			Small Rifle			Large Pistol		Small Pistol		Shotgun	
	Standard	*Magnum*	*Benchrest*	*Standard*	*Magnum*	*Benchrest*	*Standard*	*Magnum*	*Standard*	*Magnum*	*10, 12, 16, 20, 28*	*.410*
Alcan Max-Fire	Max-Fire LR			Max-Fire SR			Max-Fire LP		Max-Fire SP			
CCI	200	250	BR-2	400	450	BR-4	300	350	500	550	109, 209 T&S, 209M	
CIL	8½	8½M		1½			2½		1			
Federal	210	215	210M	200	205	205M	150	155	100		209	410
Fiocchi											209	
Remington	9½	9½M		6½	7½		2½		1½	5½	(No. 97) 209	
RWS	5341	5333		4033			5337		4031	4047		
Winchester	8½–120 WLR			6½–116 WSR			7–111 WLP		1½–108 WSP	1½M–108 WSPM	209	

groups. Hunters may or may not find better results with CCI's Benchrest primers. You have to try them before you'll know, but switching to Benchrest primers does not automatically mean loads will be more accurate.

For the varmint hunter or anyone interested in a high degree of accuracy, it is important to check a rifle's zero whenever changing brands or types of primers. If the load remains exactly the same except for a primer change, the load will usually shoot close to the same point. Against deer-sized targets, you probably will not notice much difference. On woodchucks, rockchucks, crows, or prairie dogs, a zero shift on an inch or even less might mean a miss, so it will pay to resight the rifle when there are any changes in the load.

Occasionally, one brand or type of primer will outshoot all others with a certain type of powder. The difference may be slight, certainly not as marked as switching powders, but it can be noticed. If the ultimate in accuracy is your goal, and all other things are in order, some experimenting with primers can be invaluable.

Shotshell primers are larger and heavier than rifle or pistol primers—larger because they must reach through the base wad to deliver their spark. Because the base wad, which surrounds the primer, is made of paper or plastic, the shotshell primer has a copper plated steel sleeve around it for support, called a battery cup. The battery cup holds the cup of explosive compound (cap) and the larger anvil, and has a flash hole in the top.

For years, American shotshell primers were made in two sizes: the Remington 57 and those patterned after the Winchester 209. The latter, with a diameter of .243 inch, was the most popular. Remington's 57 was slightly smaller and was accepted only by Remington cases. All other cases

manufactured in this country were designed for the 209 primer. Just why Remington chose to march to a different drummer is anyone's guess, but eventually they fell into step with the rest of the world, adopted a 209-sized primer themselves, and opened up the primer pockets on their cases accordingly.

Federal shotshell primers are produced in two types, designated 209 and 410. Both are the same size but the 410 primer is designed to ignite smaller charges of powder like those found in 410- and 28-gauge loads. The 209 primer is meant for 10-, 12-, 16-, and 20- gauge shells.

CCI discontinued its No. 157 "Remington" primer about seven years ago. The firm now offers a No. 209 Trap and Skeet, a 209M (magnum) and a 109 (designed as a hunting primer) in its line.

Primers should be stored in their factory containers. Stories about the guy who dumps all his primers into a bottle and his resultant trouble illustrate how friction, static electricity, or even slippery fingers can create hazards when primers are stored improperly. Besides, you can always tell what type of primers you have if you leave them in their boxes. And they should be kept in a cool, dry place. Exposure to high temperatures (100° or more) for an extended period of time can cause the explosive compound to deteriorate.

As reloaders, we make use of modern technology every time we assemble custom ammunition. Technology provides us with some of the most refined, consistent products ever developed. Primers are one of the most used, highly developed components available to the shooting public, and they have helped make the hobby of reloading a fascinating one, indeed. Who says reloading is only to save money?

Carmichel's Guide to Hunting Cartridges

Jim Carmichel

The debate never ends. Knock-down power, bullet energy—what are they? How do we compare one cartridge or bullet to another and determine which is better for killing big game?

Ever since hunters first observed the awesome effect of a powder-driven projectile, they've never tired of speculating on the relative effectiveness of bullets of different sizes and weights traveling at different speeds. The very earliest firearms literature, as you might guess, is filled with all sorts of opinions about killing energy. The speculations are based on everything from black magic to the alignment of stars and other heavenly bodies.

During this century, there have been some notable attempts to categorize cartridge performance. One much-discussed example is John Taylor's tables of "Knock Out," or "KO," values, in which he attempted to index the relative punch of dangerous-game cartridges.

Ballisticians prefer the foot-pound system for comparing cartridge performance for some very good reasons. First of all, it is the industry standard of comparison. Second, it is easily calculated. Third, the relative energy levels of various cartridges tend to correspond to actual performance in the field. The foot-pound energy index is not without a host of critics, however, especially shooters who prefer large-diameter, slow-moving bullets. As these critics very correctly point out, the foot-pound system is heavily skewed in favor of high-velocity cartridges. With the accepted formula for calculating bullet energy, if you double the weight of the bullet, leaving all else the same, you merely double the energy level. But if you double the velocity, you *quadruple* the energy. That's why big-bore fanciers stomp and sputter and dream up calculations that favor heavy bullets. One such set of calculations that I recently reviewed led to the conclusion that a thrown brick would be the surest way to stop a charging Cape buffalo!

A once-popular means of comparing cartridge effectiveness that was favored by fans of large, heavy bullets was the so-called, and often confusing, "pounds-feet" theory. Obviously, one reason it was so confusing was that it was frequently confused with foot-pounds of kinetic energy because of the similarity of names. The pounds-feet theory does not favor velocity as much as kinetic-energy calculations do. For example, when using standard kinetic foot-pounds calculations, a 180-grain bullet from a .30–06 leaving the muzzle at 2,700 feet per second develops 2,913 foot-pounds of energy at the muzzle. By comparison, a 405-grain .45–70 bullet that leaves the muzzle at 1,330 fps has only 1,590 foot-pounds of energy. But if we apply the pounds-feet theory, the .45–70 has a relative index of 76.95 compared with 69.40 for the .30–06.

The pounds-feet theory was done asunder by the arrival of modern cartridges, which demonstrated that the shock generated by a relatively small but high-velocity bullet was a better killer than the big holes made by old-fashioned, slow-moving bullets.

Back in the early 1960s, P.O. Ackley, the well-known gunsmith and cartridge experimenter, wrote a book called *Handbook for Shooters & Reloaders*, which includes a chapter on killing power. Ackley discusses the various theories of bullet energy, much as I have done in this article, and also includes a section by Paul Van Rosenberg on the relative effectiveness of different cartridges on big game. Ackley describes Van Rosenberg as an experienced big-game hunter and ballistics engineer, and Van Rosenberg's comments indicate

This article first appeared in Outdoor Life

Stanley W. Trzoniec photo

POINT-BLANK TRAJECTORIES AND REMAINING ENERGY LEVELS OF 100 POPULAR RIFLE CARTRIDGES

Caliber	Bullet Weight (Grains)	Bullet Type*	Muzzle Velocity (Feet Per Second)	Trajectory (inches) 50 Yards	100 Yards	150 Yards	200 Yards	250 Yards	300 Yards	350 Yards	Range at Which Bullet Is Three Inches Low	1,200 Foot-Pounds (Deer, Antelope)	2,000 Foot-Pounds (Elk, Bears to 600 Pounds)	2,800 Foot-Pounds (Large Bears, Moose)
.223 Remington	55	HPPL	3,240	1.0	2.6	3.0	2.1	0.4	4.8		282	0	0	0
.22/250 Remington	55	HPPL	3,730	0.7	2.3	3.0	2.7	1.4	1.2	5.3	324	100	0	0
.224 Weatherby Mag.	55	PE	3,650	0.8	2.3	3.0	2.7	1.3	1.4	5.5	322	120	0	0
.220 Swift	52	HPST	4,000	0.6	2.1	2.9	2.9	2.1	0.3	2.6	355	170	0	0
.243 Winchester	80	HPPL	3,350	0.9	2.5	3.0	2.4	0.5	2.8	7.9	302	200	0	0
.243 Winchester	100	PSPCL	2,960	1.1	2.6	3.0	2.0	0.5	4.5		284	250	0	0
6mm Remington	80	HPPL	3,470	0.9	2.4	3.0	2.6	0.9	2.1	6.7	311	220	0	0
6mm Remington	100	PSPCL	3,130	1.0	2.5	3.0	2.3	0.3	3.1		299	300	0	0
.240 Weatherby Mag.	87	PE	3,500	0.8	2.3	3.0	2.7	1.3	−1.3	5.2	324	320	85	0
.240 Weatherby Mag.	100	PE	3,395	0.8	2.4	3.0	2.6	1.2	−1.4	5.3	322	415	140	0
.250 Savage	87	PSP	3,030	1.1	2.6	3.0	1.9	−0.8	−5.3		278	155	0	0
.250 Savage	100	PSP	2,820	1.2	2.7	2.9	1.4	−1.8	−7.1		263	160	0	0
.257 Roberts	87	PSP	3,170	1.0	2.6	3.0	2.2	−0.1	−4.0		289	190	0	0
.257 Roberts	117	SPCL	2,650	1.4	2.9	2.7	0.5	−3.8			140	0	0	
.25/06 Remington	87	HPPL	3,440	0.9	2.4	3.0	2.5	0.6	−2.7	−7.9	303	225	50	0
.25/06 Remington	100	PSPCL	3,230	1.0	2.5	3.0	2.3	0.4	−3.1		299	285	65	0
.25/06 Remington	120	PSPCL	3,010	1.1	2.6	3.0	2.1	−0.2	−4.0		289	360	95	0
.257 Weatherby Mag.	87	PE	3,825	0.7	2.1	2.9	2.9	2.0	0.0	−3.0	351	415	175	5
.257 Weatherby Mag.	100	PE	3,555	0.8	2.3	3.0	2.8	1.6	−0.7		335	455	190	1
.257 Weatherby Mag.	117	NP	3,300	0.9	2.4	3.0	2.5	0.9	−1.9	−6.1	315	465	195	5
6.5 mm Remington Mag.	120	PSPCL	3,210	1.0	2.5	3.0	2.4	0.5	−2.9	−7.7	302	390	155	0
.264 Winchester Mag.	100	PSP	3,320	0.9	2.5	3.0	2.4	0.4	−3.1		299	270	80	0
.264 Winchester Mag.	140	PSPCL	3,030	1.1	2.6	3.0	2.2	0.0	−3.6		299	475	205	10
.270 Winchester	100	PSP	3,480	0.3	2.4	3.0	2.6	0.9	−2.1	−6.7	311	295	85	0
.270 Winchester	130	BP	3,110	1.0	2.5	3.0	2.3	0.3	−3.1		299	435	170	0
.270 Winchester	150	SPCL	2,900	1.2	2.7	2.9	1.6	−1.6	−6.7		266	295	115	15
.270 Weatherby Mag.	130	PE	3,375	0.8	2.4	3.0	2.6	1.2	−1.4	−5.3	323	570	295	95
.270 Weatherby Mag.	150	NP	3,245	0.9	2.4	3.0	2.5	0.9	−1.9	−6.0	316	690	380	155
.284 Winchester	125	PP(SP)	3,140	1.0	2.5	3.0	2.2	0.1	−3.5		294	370	150	0
.284 Winchester	150	PP(SP)	2,860	1.2	2.7	2.9	1.7	−1.1	−5.6		274	395	155	0
7/30 Waters	120	FP	2,680	1.4	2.9	2.7	6.5	−3.9	−11.2		242	145	0	0
7mm Mauser	140	PSP	2,660	1.3	2.8	2.8	1.2	−2.1	−7.4		260	325	55	0
7mm/08 Remington	140	PSP	2,860	1.2	2.7	2.9	1.8	−0.8	−5.1		278	410	140	0
.280 Remington	150	PSPCL	2,970	1.1	2.6	3.0	2.0	−0.5	−4.5		284	435	195	25
.280 Remington	165	SPCL	2,820	1.2	2.7	2.9	1.4	−1.8	−7.0		264	355	160	15
7mm Remington Mag.	125	PP(SP)	3,310	0.9	2.4	3.0	2.5	0.7	−2.5	−7.1	307	410	195	40
7mm Remington Mag.	150	PSPCL	3,110	1.0	2.5	3.0	2.2	0.2	−3.4	8.5	296	485	245	75
7mm Remington Mag.	175	PSPCL	2,860	1.2	2.7	2.9	1.9	−0.7	−4.8		281	575	285	80
7 mm Weatherby Mag.	139	PE	3,300	0.9	2.4	3.0	2.5	1.0	−1.8	−6.0	316	570	300	110
7 mm Weatherby Mag.	154	PE	3,160	1.0	2.5	3.0	2.4	0.6	−2.4		307	625	340	130
7 mm Weatherby Mag.	175	RN	3,070	1.0	2.5	3.0	2.3	0.3	−3.0	−7.8	300	660	380	175
.30 Carbine	110	SP	1,990	2.1	2.9	0.1	−7.1				176	0	0	0
.30 Remington	170	ST	2,120	1.9	3.0	1.5	−3.1				199	115	0	0
.30/30 Winchester	150	SPCL	2,390	1.7	3.0	2.1	−1.6	−8.7			212	120	0	0
.30/30 Winchester	170	SPCL	2,200	1.8	3.0	1.8	−2.3	−9.6			206	140	0	0
.300 Savage	150	SPCL	2,630	1.4	2.9	2.6	0.3	−4.4			238	280	65	0
.300 Savage	180	PSPCL	2,350	1.6	2.9	2.4	−0.2	−5.2			231	305	50	0
.300 Savage	180	RM	2,350	1.7	2.3	2.2	−1.1	−7.3			218	300	35	0
.30/40 Krag	180	PSPCL	2,430	1.5	2.9	2.5	0.2	−4.3			238	345	90	0

*HPPL, Hollow Point-Lokt; PE, Pointed Expanding; HPBT, Hollow Point Boat Tail; PSPCL, Pointed Soft Point Core-Lokt; PSP, Pointed Soft Point; SPCL, Soft Point Core-Lokt; NP, Nosier Partition; BP, Bronze Point; PP(SP), Power-Point (Soft Point); FP, Flat Point; RN, Round Nose; SP, Soft Point; ST, Silver Tip; SJHP, Semijacketed Hollow Point; FMJ, Full Metal Jacket.

				Trajectory (inches)							Ranges at Which Cartridges Retain Three Levels of Energy (Yards)			
Caliber	Bullet Weight (Grains)	Bullet Type*	Muzzle Velocity (Feet Per Second)	50 Yards	100 Yards	150 Yards	200 Yards	250 Yards	300 Yards	350 Yards	Range at Which Bullet Is Three Inches Low	1,200 Foot-Pounds (Deer, Antelope)	2,000 Foot-Pounds (Elk, Bears to 600 Pounds)	2,800 Foot-Pounds (Large Bears, Moose)
.30/40 Krag	220	ST	2,160	1.8	3.0	1.9	−1.7	−8.1			214	330	85	0
.30/06 Springfield	110	PSP	3,380	1.0	2.5	3.0	2.2	−0.2	−4.5		285	225	85	0
.30/06 Springfield	150	BP	2,910	1.1	2.7	2.9	1.9	−0.7	−4.9		280	435	185	5
.30/06 Springfield	165	PSPCL	2,800	1.2	2.7	2.9	1.5	−1.5	−6.4		268	405	180	15
.30/06 Springfield	180	PSPCL	2,700	1.3	2.8	2.8	1.3	1.9	7.0		263	460	205	20
.30/06 Springfield	220	SPCL	2,410	1.6	2.9	2.4	−0.3	5.5			229	330	140	5
.300 Winchester Mag.	150	PSPCL	3,290	0.9	2.5	3.0	2.4	0.6	2.7	7.5	304	465	260	115
.300 Winchester Mag.	180	PSPCL	2,950	1.1	2.6	3.0	2.1	−0.1	3.8		291	645	355	145
.300 Winchester Mag.	220	ST	2,680	1.3	2.8	2.8	1.3	2.0	7.3		261	540	300	125
.300 H&H Mag.	150	ST	3,130	1.0	2.6	3.0	2.2	0.1	3.6		294	460	245	90
.300 H&H Mag	180	PSPCL	2,880	1.2	2.7	2.9	1.8	−0.7	5.0		279	535	280	95
.300 H&H Mag.	220	ST	2,580	1.4	2.9	2.7	0.8	−3.0	9.0		250	490	260	95
.300 Weatherby Mag.	150	PE	3,545	0.8	2.3	3.0	2.7	1.5	−0.8	4.3	334	645	400	225
.300 Weatherby Mag.	180	PE	3,245	0.9	2.4	3.0	2.5	0.9	−1.9	6.0	315	750	465	265
.303 Savage	190	ST	1,940	2.1	2.9	0.6	−5.4				184	70	0	0
.307 Winchester	150	FP	2,760	1.3	2.8	2.7	0.6	3.8	−11.1		243	205	70	0
.307 Winchester	180	FP	2,510	1.5	2.9	2.5	0.0	−5.1	−13.0		232	250	80	0
.303 British	180	SPCL	2,460	1.6	2.9	2.4	−0.4	−5.8			227	230	65	0
.308 Winchester	110	PSP	3,180	1.1	2.6	3.0	1.8	−1.1	−6.3		271	200	60	0
.308 Winchester	125	PSP	3,050	1.1	2.6	3.0	1.9	−0.6	−5.0		280	295	105	0
.308 Winchester	150	PSPCL	2,820	1.2	2.7	2.9	1.5	−1.5	−6.5		267	345	130	0
.308 Winchester	180	PSPCL	2,620	1.4	2.8	2.8	1.1	−2.5	−8.1		256	425	170	0
.32 Winchester Special	170	SPCL	2,250	1.8	3.0	1.9	−2.0	−9.2			209	145	0	0
8mm Mauser	170	SPCL	2,360	1.7	3.0	2.0	−1.6	−8.6			212	150	15	0
8mm Remington Mag.	185	PSPCL	3,080	1.1	2.6	3.0	2.1	−0.2	−4.1		287	490	290	150
8mm Remington Mag.	220	PSPCL	2,830	1.2	2.7	2.9	1.7	−1.1	−5.7		273	580	345	180
.338 Winchester Mag.	200	PP(SP)	2,960	1.1	2.7	3.0	1.9	−0.8	−5.1		278	495	295	150
.338 Winchester Mag.	225	SP	2,780	1.2	2.7	2.9	1.7	−1.1	−5.6		274	680	400	205
.338 Winchester Mag	250	ST	2,660	1.3	2.8	2.8	1.0	2.6	8.3		255	700	320	175
.340 Weatherby Mag.	200	PE	3,210	0.9	2.5	3.0	2.4	0.7	−2.4	−6.9	308	695	455	280
.340 Weatherby Mag.	250	NP	2,850	1.2	2.7	2.9	1.6	−1.3	−6.1		270	530	360	220
.35 Remington	150	PSPCL	2,300	1.8	3.0	1.8	−2.5	−10.8			203	95	0	0
.35 Remington	200	SPCL	2,080	2.0	3.0	1.0	−4.8				188	115	0	0
.351 Remington SL	180	RN	1,850	2.2	2.8	−0.1	−7.3	−19.6			174	40	0	0
.356 Winchester	200	FP	2,460	1.6	2.9	2.4	−0.5	−6.0	−14.8		226	255	95	0
.356 Winchester	250	FP	1,160	1.8	3.0	1.8	−2.1	−9.0	−19.5		208	300	105	0
.358 Winchester	200	ST	2,490	1.5	2.9	2.5	−0.1	−5.1			232	290	115	0
.358 Winchester	250	ST	2,230	1.7	3.0	2.1	−1.4	−7.6			216	355	105	0
.350 Remington Mag.	200	PSPCL	2,710	1.3	2.8	2.8	1.1	−2.5	−8.3		255	395	205	65
.375 Winchester	200	FP	2,200	1.8	3.0	1.6	−2.9	−11.1	−23.9		201	160	20	0
.375 Winchester	250	FP	1,900	2.1	2.9	0.5	−5.6	−16.0	−31.3		183	175	2	0
.375 H&H Mag.	270	SP	2,690	1.3	2.8	2.8	1.1	−2.4	−8.0		257	550	345	200
.378 Weatherby Mag.	300	NP	2,925	1.2	2.7	2.9	1.8	−1.0	−5.5		275	640	440	310
.38/40 Winchester	180	SP	1,160	3.0	−0.2	−12.1					116	0	0	0
.38/55 Winchester	255	SP	1,320	2.8	1.5	−5.9					135	0	0	0
.44 Remington Mag.	240	SJHP	1,760	2.4	2.6	−1.7	−11.8				159	65	0	0
.444 Marlin	240	SP	2,350	1.8	3.0	1.6	−3.3				197	170	80	0
.444 Marlin	265	SP	2,120	1.9	3.0	1.1	−4.3				191	190	70	0
.45/70 Government	405	SP	1,330	2.8	1.4	−6.3					134	110	0	0
.458 Winchester Mag.	500	FMJ	2,040	1.9	3.0	1.5	−3.0	−10.8			200	640	360	220
.460 Weatherby Mag.	500	FMJ	2,700	1.3	2.8	2.8	1.1	−2.4	−8.0		257	827	575	445

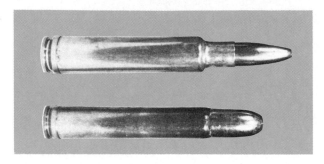

Which is more effective on dangerous game? Bullet weight or bullet velocity? Cartridges shown offer the two extremes. At top is the .378 Weatherby Magnum, which delivers 5700 foot-pounds energy with 300 grain bullet and a muzzle velocity of 2925 f.p.s. The cartridge below it is the .458 Winchester Magnum, which delivers 4620 foot-pounds energy with a 500 grain bullet at a comparatively slow 2040 f.p.s.

this to indeed have been the case. Among Van Rosenberg's more interesting recommendations is the establishment of more-or-less specific energy levels that a cartridge must provide to perform well on game. He feels that 1,200 foot-pounds of energy is adequate for game such as deer, antelope, sheep, and goats. For elk and small bears, he recommends 2,000 foot-pounds as adequate, and he sets 2,800 as the adequate energy level for large bears and moose. Of course, Van Rosenberg speaks of the *remaining energy at the target*, not at the muzzle.

By necessity, Van Rosenberg's recommendations must be regarded as generalizations, and there are plenty of exceptions to them. For example, the energy level of the .30–30 Winchester with a 170-grain bullet drops below 1,200 foot-pounds inside 150 yards. Yet, we know by experience that a .30–30 easily kills deer at 200 yards and even a bit beyond. Just the same, we can make some interesting comparisons of different cartridges and bullet weights by imposing Van Rosenberg's performance levels and seeing how— or, rather, where (at what ranges)—they stack up.

The accompanying table, computed at *Outdoor Life's* Briarbank Ballistic Laboratory, shows the critical ranges for 100 cartridges and bullet weights. By comparing the different energy/range figures, you'll get a pretty fair idea of what to expect in the way of *relative* performance on big game at different ranges.

Also included is an expanded version of "The Nonthinking Man's Trajectory Table." This proved mighty popular when *Outdoor Life* printed it a couple of years ago. The table shows the three-inch (plus or minus) point-blank range of most American big-game calibers. By three-inch point-blank range, we mean that, within the recommended ranges, the bullet never rises above or

falls below three inches of your line of sight. From the practical hunter's standpoint, this is more than adequate bullet placement because it is well within the vital-area size of all big-game animals and many varmint species.

These figures are the best-ever means of comparing the useful hunting ranges of various cartridges. All data are based on a line of sight 1½ inches above the bore line. This is typical of most scope-sighted rifles. To make use of the table, simply sight your rifle in at any of the ranges shown so that the bullet impact, in relation to point of aim, matches the impact point at that distance (assuming that you're using factory-loaded ammunition and a rifle that is in good condition).

For example, the table shows that the .270 Winchester with a 130-grain Bronze Point bullet is 2.5 inches high at 100 yards. Accordingly, simply adjust the sight on your .270 so that the bullet hits 2.5 inches above point of aim at 100 yards. From that point on, the bullet will hit within three inches (vertically) of where you aim out to more than 250 yards.

One column of the table lists the range at which the bullet is three inches below line of sight. All the way out to that range, stated in yards, you can hold dead on and forget about trajectory when you're big-game hunting or when you're shooting most varmints. What could be more practical?

By consulting the retained-energy figures and the point-blank ranges, you can provide yourself with a very good idea of the performance of any listed cartridge. Some have such a looped trajectory that the practical hunting ranges at which they can be used are very short. At the other extreme are some of the magnums that retain, for instance, enough energy to kill deer dead at very long ranges. But at those ranges (over 300 yards), even expert riflemen would seldom attempt a shot. In other words, by using both sets of figures, you can quite closely determine how far you can kill efficiently in terms of both on-target energy and ability to make a hit. Perhaps best of all, the combined figures give someone who's buying a new rifle a reliable means of determining which cartridge or cartridges are best suited to his form of hunting.

Hitting what you shoot at is the big weakness in the assumptions behind most cartridge-performance tables. Argue as we may about the relative performance of various cartridges, the major factor is the man behind the gun. When it comes to killing game, I give bullet placement a relative importance of 70 percent. The other 30 percent can be divided between bullet energy and terminal bullet performance anyway you want to split it up. As for myself, I long ago decided not to agree with anyone.

Does the Wad
Make a Difference?

Hugh Birnbaum

Many trapshooters who reload are somewhat casual about their wad choice. This is probably because the most widely used reloading manuals, in a laudable effort to present as much helpful data as possible, sometimes list a bewildering array of component combinations for the more popular shotshells. As an example, the *Reloaders' Guide for Hercules Smokeless Powders* offers load specifications for 10 different wads deemed suitable for assembling 3-dram-equivalent 1⅛-ounce trap loads with Red Dot powder in the Winchester AA-type case.

Some reloaders restrict their choices to wads manufactured by the maker of the shotshell, working on the assumption that the factory probably knows best. Others, myself included, sometimes turn to wads from a variety of manufacturers, limited only by the prerequisite that the component combination be listed in at least one authoritative reloading manual. Reasons for straying from the factory fold include the quest for the perfect pattern, temporary unavail-

ability of a preferred wad, or a dealer's supersale presenting an irresistible bargain.

More often than not, we judge the effectiveness of a load by watching targets break (or sometimes not break). Smoke is good. Chips and splits are bad. Rarely do we take the time and trouble to pattern the load formally. And so we occasionally end up blaming a good load because of poor shooting or, less often, blaming ourselves for shooting erratically while actually being sabotaged by a combination of components poorly matched to the specific gun barrel, choke tube, or shooting distance.

Recent experiences while practicing with two load combinations that were essentially identical except for the wad piqued my curiosity sufficiently to move me to undertake two grueling stints at the pattern board. The first load tended to ink-blot targets from the 16-yard line and from medium handicap yardage as well. The second

This article first appeared in American Rifleman

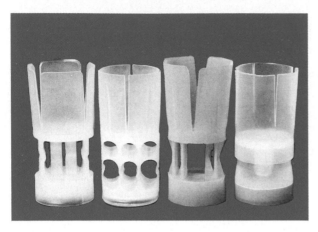

Wads used in pattern testing were (from left): Winchester WAA12, Remington RXP 12, Hornady/Pacific Versalite, and Federal 12C1.

load, differing only in the wad and 0.5 grain more powder, chipped and chunked targets, providing few soul-satisfying hits.

Innate sloth notwithstanding, I wanted to see exactly what was happening, and only pattern sheets would tell. Once committed to that drastic step, I decided to add two more wads to the comparison shooting to find out how much difference, if any, the choice of wad might make when all other factors remain virtually constant.

The loads I compared are all based on the ubiquitous Winchester AA Plus case. A friend contributed a supply of once-fired hulls from the same production lot, which I reprimed with Winchester 209 primers from a single tray of 100. The loads, all listed in the *Reloaders' Guide for Hercules Smokeless Powders*, consisted of 10 rounds each as follows:

Winchester WAA12 Wad
18.0 grains Red Dot
10,400 approximately psi

Remington RXP 12 Wad
18.5 grains Red Dot
9,800 approximately psi

Hornady/Pacific Versalite Wad
19.0 grains Red Dot
9,700 approximately psi

Federal 12C1 Wad
18.5 grains Red Dot
9,700 approximately psi

These loads, containing 1⅛ ounces of Lawrence Magnum No. 7½ high-antimony shot, are all classed as 3-dram-equivalent shotshells with a nominal velocity of 1,200 fps. I weighed all pow-

der and shot charges on an RCBS scale with a claimed accuracy of ±0.1 grain. I also counted the pellets in each payload. The grand total for the 40 shells was 15,702.

Note that the 19.0-grain powder charge the Hercules manual recommends with the Versalite wad is 1.0 grain heavier than the 18.0-grain charge Hornady suggests for the same components on the data sheet enclosed in the bag of wads. This sort of gentlemen's disagreement often crops up when comparing load data from different sources, as different test equipment, procedures, technicians, and interpretations can yield slightly different numbers. If you feel better with 18.0 grains, your targets will still break. I assembled the test loads with a MEC 600 Jr. single-stage reloading press.

The wads selected for the test shells were plucked at random from their respective containers, inspected for disqualifying defects (I found none) and weighed. The WAA12, RXP 12 and Versalite wads are of one-piece construction, each molded from a single blob of plastic. The 12C1 is a comparatively rare bird in that it consists of two separate plastic sections that friction-fit to form a single unit that can be handled as easily as the one-piece wads. All the wads entered the shells easily and compressed sufficiently to permit neat crimping without undue effort or any tendency to force the crimp open during storage.

I did the pattern shooting at North Jersey Gun Club in Fairfield, New Jersey, on two successive days. Temperatures ranged between 85 and 95

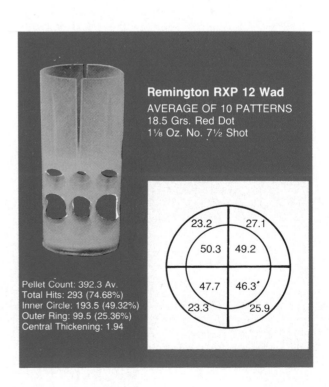

Remington RXP 12 Wad
AVERAGE OF 10 PATTERNS
18.5 Grs. Red Dot
1⅛ Oz. No. 7½ Shot

23.2	27.1
50.3	49.2
47.7	46.3*
23.3	25.9

Pellet Count: 392.3 Av.
Total Hits: 293 (74.68%)
Inner Circle: 193.5 (49.32%)
Outer Ring: 99.5 (25.36%)
Central Thickening: 1.94

degrees Fahrenheit in light breezes judged to have little or no effect on the patterns. In case you're wondering, pattern shooting at 95 degrees is not pleasant, and I do not recommend it. Patterning distance was the standard 40 yards.

The test gun was my Remington Model 1100 TB autoloader, equipped with a nearly new Remington factory-standard 28-inch Full-choke trap barrel and a Timney all-steel pull trigger that is almost riflelike in its crispness and consistency. I had put about 1,500 rounds through the barrel prior to the patterning sessions, shooting a variety of reloads. It is the least finicky trap barrel I have ever used with respect to shot size, composition, and payload weight, generally producing gratifying solid hits with everything from 1⅛-ounce loads of extra-hard No. 7½ shot to 1-ounce loads of pot-luck reclaimed shot. If I may indulge in anthropomorphism for a moment, this barrel has a decidedly friendly personality.

None of the loads generated uncomfortable felt recoil, and none felt significantly lighter or heavier than the others. Calculations of free recoil energy, assuming a gun weight of 8.5 pounds, pegged the loads at 18.88 foot-pounds with the WAA12 wad, 18.78 foot-pounds for the RXP 12 recipe, 18.99 foot-pounds for the Versalite version, and 18.77 foot-pounds with the 12C1 wad. No shoulder could detect such minute differences. All loads cycled the Model 1100's action smartly but smoothly. There were no instances of weak ejection and no sensation of violent

One test shell, containing a Versalite wad, split when fired, apparently because of weak or brittle plastic. No adverse effects on shooter, gun, or pattern resulted. All test shells were once-fired cases from a single lot of factory trap loads.

operation. My experience has been that loads that harshly jolt open an autoloader will bring on a plague of mechanical problems. Gentle is better.

Somewhat surprisingly, given the reputation for longevity of AA-style cases, shell No. 23, which contained a Versalite wad, split upon firing. I would not have known it had I not been using a shell-catcher. The report and recoil were normal, and the action operated properly. There wasn't even a noticeable anomaly involving release of powder gases. Nonetheless, the case disclosed a dramatic-looking 1.5-inch lengthwise rupture line from just below the crimp to about ⅓-inch above the brass. This may have been symptomatic of a plastic problem, as other cases from the same batch displayed premature incipient cracking and splitting along crimp folds. These minor flaws occurred with all four load variations.

It's unfortunate that editorial reality precludes publishing photographs of all the patterns, because in this case several significant differences are not apparent in the numerical data and diagrams used to condense patterning results. Before describing the patterns in detail, however, I wish to stress that the patterns I shot are specific to the barrel I used and the particular variables and conditions that affected the test

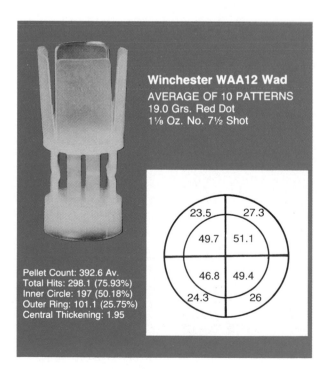

Winchester WAA12 Wad
AVERAGE OF 10 PATTERNS
19.0 Grs. Red Dot
1⅛ Oz. No. 7½ Shot

23.5	27.3
49.7	51.1
46.8	49.4
24.3	26

Pellet Count: 392.6 Av.
Total Hits: 298.1 (75.93%)
Inner Circle: 197 (50.18%)
Outer Ring: 101.1 (25.75%)
Central Thickening: 1.95

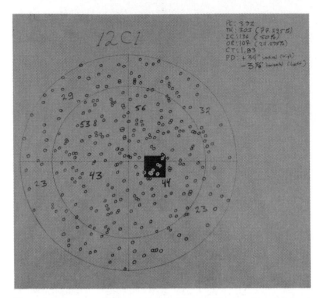

Test loads containing Federal's 12C1 wad produced beautifully even patterns, with excellent coverage of the outer ring. These reloads were clearly very compatible with the test barrel.

usually decently uniform about the periphery of the inner circle, but the outer half of the outer ring was often too sparsely populated to count on. These patterns confirmed my practical shooting with this combination of components, which I have found splendid for handicap events but too unforgiving for my sloppy pointing at 16 yards. My impression is that one must be a really good trapshooter to hang black clouds in the air every time with this load at short range with a tight choke. An Improved-Modified choke, though, makes shooting this combination a delight.

Loads containing the Versalite wad produced a 69.87 percent pattern average, slipping by a hair from Full-choke territory into the top of the Improved-Modified range. The tightest pattern yielded a solidly Full-choke 74.87 percent, and the loosest, at 64.29 percent, fell at the high end of Modified-choke performance. The pattern sheets practically shouted out loud that the test barrel didn't like this load one bit. Pellet distribution was depressingly patchy. I cannot explain this unexpectedly dismal turn of events, as I have used Versalite wads with considerable satisfaction in other barrels and other types of cases. This experience goes far to strengthen the often made but seldom followed suggestion that trapshooters pattern-test new loads in the barrels they normally use before stepping to the line with them.

Patterns produced with loads containing the Federal 12C1 wad provided a pleasant surprise.

process. Your barrel and test conditions might well yield substantially different results.

For openers, my patterns with the WAA12 wad appear to confirm the notion that one stands a good chance of obtaining optimum results by matching wad and case along brand lines. On average these loads yielded the tightest patterns, scoring a cumulative 75.93 percent total pellet hits in the 30-inch circle. One shell produced the tightest of all 40 patterns, a rousing 83.46 percent. On the other hand, another in the same series yielded only 66.75 percent, the second weakest pattern of the 40. The nearly 17 percent difference between the strongest and the weakest pattern in this group was appreciably greater than with the other loads. Nonetheless, pellet distribution was generally very good, and the outer rings tended to be well covered. According to the pattern sheets and practical clay bird crunching this load performs well in 16-yard and handicap shooting. The centers are hot enough for smokeballs and the edges are filled well enough to save the day when necessary. The test barrel clearly likes this load. So do I.

Patterns produced with the RXP 12 reloads were very nearly as tight as the WAA12 patterns, averaging 74.68 percent. However, there was less overall variation, with the tightest pattern registering 79.54 percent and the weakest 69.29 percent. Pellet distribution in the inner circle sometimes left a gap large enough to let a target escape. Pellet distribution in the outer ring was

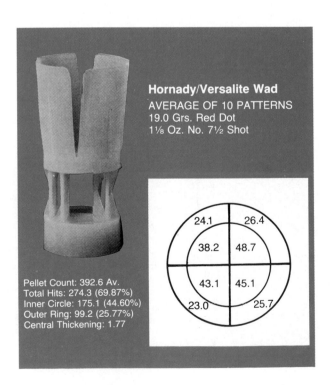

Hornady/Versalite Wad
AVERAGE OF 10 PATTERNS
19.0 Grs. Red Dot
1⅛ Oz. No. 7½ Shot

Pellet Count: 392.6 Av.
Total Hits: 274.3 (69.87%)
Inner Circle: 175.1 (44.60%)
Outer Ring: 99.2 (25.77%)
Central Thickening: 1.77

24.1	26.4
38.2	48.7
43.1	45.1
23.0	25.7

The overall pattern average was 73.69 percent, the densest pattern tallied 77.29 percent, and the weakest one a very respectable 69.64 percent. The rather low 7.65 percent spread between the tightest and loosest patterns is impressive in and of itself. Scanning the pattern sheets revealed an even rosier picture. Pellet distribution tended to be quite even, center and edge, with very few "lead fist" clumps and even fewer pellet-free patches. And there was more good news. The patterns boasted genuinely useful outer rings. To a shooter like me, who needs good edge coverage to help compensate for innate lack of talent combined with all too frequent lapses of concentration, this load seems made to order.

The reason the fine performance of the 12C1 load surprised me is that the 12C1 wad is primarily intended for use with the Federal Champion paper case, which is straight-walled and cavernous. It is about as different from the compression-formed plastic Winchester AA Plus case as can be. I knew that the combination of 12C1 wad and AA Plus case was safe, because the Hercules reloading manual would not have listed it if it weren't. And I certainly would not have spent considerable time and effort patterning a load I expected to perform poorly. I'm neither a sadist nor a masochist. In any case, I was delighted to strike gold unexpectedly with this somewhat unorthodox mix of components.

The bottom line, after counting and analyzing 11,548 pellet holes on 40 pattern sheets, is that the choice of wad can in fact have a profound

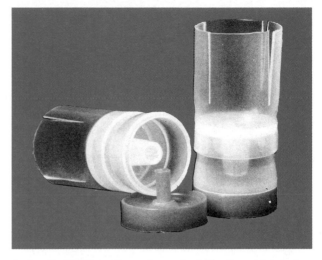

Federal's 12C1 is the only wad of the four tested that uses a two-piece construction. The two parts are packaged as a unit.

WAD WEIGHTS COMPARED				
	Win. WAA12, grs.	Rem. RXP 12, grs.	Hornady/ Pac. Vers., grs.	Fed. 12C1, grs.
Lightest	40.0	36.3	38.0	37.0
Heaviest	41.8	39.0	39.7	37.7
Ext. Sp.	1.8	2.7	1.7	0.7
Average	40.78	37.58	39.29	37.41

The author weighed 10 randomly selected samples of each wad on an RCBS reloading scale with a claimed accuracy of ±0.1 gr. Surprisingly, in view of their complex construction, the 12C1 wads exhibited the greatest weight uniformity.

effect on the quality of the pattern. It may not show up in numerical pattern synopses, but it may be obvious when viewing the pattern sheets. I used the word "may" in the preceding sentence because it is likely that some barrels and/or choke tubes are more sensitive than others to wad differences. I am sure that somewhere there is a trapshooter with a barrel that handles all types of wads indistinguishably. And I know for a fact that I have one barrel that reacts quite differently according to the wad used when shooting reloads in AA Plus cases as described above.

Given the perverse nature of shotgun barrels in general and the variables possible in reloading, I wouldn't care to make a more sweeping statement. Except perhaps that you'll never know for sure what to expect of a given load unless you actually pattern it in your own gun. It's a pain, I know. But who ever said having fun was supposed to be easy?

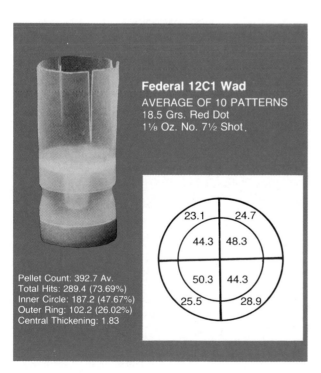

Federal 12C1 Wad
AVERAGE OF 10 PATTERNS
18.5 Grs. Red Dot
1⅛ Oz. No. 7½ Shot.

23.1	24.7
44.3	48.3
50.3	44.3
25.5	28.9

Pellet Count: 392.7 Av.
Total Hits: 289.4 (73.69%)
Inner Circle: 187.2 (47.67%)
Outer Ring: 102.2 (26.02%)
Central Thickening: 1.83

Handloading Secrets for Big-Game Hunting

Bob Milek

Handloads to be used for big-game hunting deserve as much attention as loads intended for varmint shooting or target work. Ironically, a large segment of the shooting public seems to feel that just any old load will suffice for big game. To some extent this is true, because it doesn't really take anything extra-special to put down that buck. However, it's not just that your handload is capable of downing a buck that counts. There's a special thrill in filling your tag using a load that you spend long hours tailoring just for your rifle. There are side benefits as well. You get a lot of practice shooting while you're searching for that "best" load and you'd be surprised knowing exactly what your load will do at various ranges does for your confidence when that one-in-a-season chance at a trophy comes along.

I suspect that many hunters fail to tailor loads for their big-game rifle for one of two reasons—either they don't know how to go about it or they think the job takes too much time. As for the latter, you bet it takes time. But can you think of a better way to fill idle hours in the summer months preceding the fall hunting season? As far as knowing how to go about tailoring a load for your big-game rifle, it's simple once you sit down and give the matter some thought.

Consider my most recent load development job. The 1984 big-game season was coming up and I wanted a load for my Thompson/Center T/CR 83 .30–06 that I could use for both mule deer and elk here in Wyoming. My first step was to decide on what bullet weight I wanted to use. A 150-grainer would work fine on mule deer, but it's a bit light for elk if a lot of bone has to be broken to reach the vitals. Years of experience have proved to me that a good 180-grain bullet in the .30–06 is an excellent choice for elk, giving both good penetration through heavy bone and plenty of

Whether for trophy hunting or just bringing in winter meat, careful load development in an adequate caliber, such as the versatile .30–06, may make the difference in the outcome of a carefully planned hunt.

expansion to inflict quick-killing damage to tissue and the vital organs. However, a 180-grain .30 caliber bullet can be a little on the heavy side for mule deer, sometimes failing to give adequate expansion.

What I wanted was something in between, a bullet I could count on to perform on both mule deer and elk under most of the conditions I anticipated. A 165-grain .308-inch bullet—a weight many shooters consider ideal for the .30–06—was the obvious answer.

Now as to bullet form. I'd be hunting both mule deer and elk in the same area where the terrain is dominated by deep, rocky canyons, huge sage-covered basins, and draws with skimpy patches of timber found only on the highest ground. In other words, my shooting would most likely come at any distance from 100 to 300 yards or more.

This article first appeared in Guns & Ammo

Author Milek found Thompson/Center's first centerfire rifle, the T/CR 83, in .30–06, adequate in power and accuracy to anchor game at extreme ranges. He does not feel handicapped in any way with a single-shot.

For such shooting a streamlined, pointed bullet that retains its velocity well is essential. After checking my bullet shelf, I settled on five bullets to test—Hornady's 165-grain spire point and their 165-grain boattail spire point, the Nosler 165-grain solid base spitzer, the Speer 165-grain spitzer, and the Sierra 165-grain boattail hollow point. The latter has a poorer ballistic coefficient than the others, but is still a good long-range performer. It mattered not one bit to me which of the five bullets I used—they're all excellent game bullets. The one that proved to be the most accurate in my rifle would be the one I'd use on deer and elk.

Powder next. The list of those that give excellent results in the .30–06 is almost endless. But I had a prerequisite that reduced the selection considerably. Whatever powder I used had to nearly fill the case, creating a condition of high loading density which I feel frequently contributes to optimum accuracy. This rules out use of all but the slow-burners. Hodgdon's 4831 and 4350 DuPont 4350, Norma MRP, Hodgdon's H414 and H450, and Winchester 760 were the ones I finally settled on trying. The three ball powders—H414, H450, and Winchester 760—would yield the poorest loading density, but all three have shown me good performance in other .30–06 rifles, so I wanted to check them in the T/CR.

For primers I chose to use standard large rifle—

of various brands—whenever possible. Because ball powders are difficult to get started, particularly in cold weather, I'd use magnum primers with them. With the other powders, though, the .30–06 doesn't hold enough of any of them to require magnum primer heat and extended burning duration to effect complete, consistent combustion.

At this point I should point out that preselection of the bullet weight and style, as well as the powders to be tested, has a drawback. It's possible that none of the combinations will produce the optimum accuracy a rifle is capable of. Still, I think that it's important to use a bullet that will do the particular jobs I have for it and if this means settling for a little less than the best possible accuracy, so be it.

When the planning is completed, it's time to go to work. The first step is selection of cases. Here it's important to use cases of the same make. There's enough dimensional variation among cases of the same make without adding the problems of mixing brands where the capacities can vary greatly. Such variations will have an adverse effect on both pressure and accuracy, ruining otherwise well-conducted tests.

After inspecting the cases and discarding any with defects, run them all through full-length sizing dies. I want to stress *full-length resizing* as opposed to neck sizing. On a hunting trip each

Author prefers full-length sizing of cases (left), and careful trimming to length (above), for assurance of feeding and proper ejection, even in single-shot rifles.

round must chamber easily and this can be guaranteed only when you full-length resize. A neck-sized case may come much closer to fitting your chamber perfectly, but it will also be more difficult to chamber and extract. On occasion a neck-sized case won't chamber. Or, if you manage to fully chamber a tight round, it may not extract. You don't have to have much of an imagination to see what failure to chamber or extract can do to a hunting trip. Play it safe—always full-length resize!

After resizing and decapping, both accomplished in a single operation in today's dies, it's a good idea to trim all of the cases to the same length. I trimmed my .30–06 cases to 2.484 inches, .010 inch under the maximum. When you seat your bullets with a friction fit rather than crimping them in place, uniform case length is of little importance as long as no case exceeds the maximum length. Nevertheless, having them all the same length is just one of those little trivials that I feel good about.

Next you must clean the cases to remove all of the resizing lubricant. This is important! Besides collecting grit which is damaging to your rifle's chamber, lubricant left on the cases prevents the case, which expands at the moment of ignition, from momentarily grabbing the chamber walls as it should. The result is increased backthrust on the breech face which gives indications of high pressure with powder charges normally developing safe pressure. You can wipe the cases clean with a solvent-soaked rag or tumble them until they're clean. Whatever the method, clean the resized cases thoroughly.

Now it's decision time. You can load the rounds in your reloading room, then go to the range to

After sizing, bullet lube must be removed from cases by wiping with solvent or by tumbling.

test them, but this takes a lot of time. Remember, with each bullet and each powder you have to begin with the starting load shown in the reloading manual and work up in increments of no more than one-half grain until you reach the maximum for your rifle. Sticky extraction, flat primers, or best of all, case head expansion, are pressure indicators you must monitor carefully. Each load requires three rounds so that you can check accuracy. The problems in logistics when you try to do all of the loading at home, then shoot at the range, are obvious.

By far the best system is to do your reloading right at the range. This isn't nearly as difficult as it sounds. You'll need a press for bullet seating, a powder scale for weighing charges, and some means of seating primers. A powder measure is a handy extra, but not a must. A nifty tool for this work is the Huntington Compact Press, a strong, powerful, and very portable hand press on which you can seat both primers and bullets. Then, should you run out of prepared cases, you can easily full-length resize with the Huntington tool. However, lacking this handy-dandy piece of equipment, it's no big chore to use your press at

the range. Simply secure it to the shooting bench using a couple of hefty C-clamps. Of course, no matter what press you use, you'll have to set up your scale on the bench to weigh charges. A word of warning: you can work with the scale at the range only on absolutely calm days; any breeze whatsoever will work on the scale pan and you'll never get the beam to settle down and give you a correct reading.

I prefer to prime my cases separate of the press, so I use the Lee Auto-Prime tool at the range. This is a hand-operated tool that allows me to seat each primer with thumb pressure. I can feel the primer in and I know exactly when it bottoms, thus avoiding crushed primers or a high primer condition, both of which can occur when you use the priming fixture on your press.

Your first load should be the starting load shown in your reloading manual, not the middle or maximum load shown. Many shooters think they can start at the top and work up, their reasoning being that all of the manuals are conservative because of liability. *Bull!* Those loads are developed by expert ballisticians with equipment designed to monitor pressure. They know their business and you'll do well to pay attention. Sure, you may be able to safely exceed the published maximum in your particular rifle. But then, too, you may encounter high pressure long before you near the published maximum. Variations in chamber and bore dimensions, temperature, humidity, case capacity, powder burning rate, and other variables combine to change chamber pressure and load performance from rifle to rifle or load to load. It's these variables that account for the differences in loading data between one manual and another and they're the reason that you should always start low and work up carefully.

Strength and handiness of the Huntington Compact Press make it ideal for in-the-field work.

"C" clamps can be used in the field to hold press to bench, combining stability with portability.

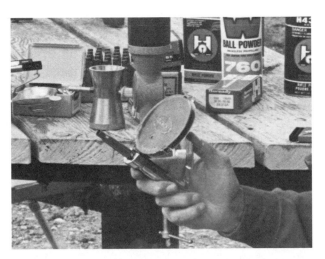

Lee's Auto-Prime tool allows the handloader to actually "feel" the primer into the primer pocket.

I prefer to pick one powder and try all bullet and charge combinations with it before moving to another powder. I load three rounds, fire them for accuracy at 100 yards, then advance the charge no more than one-half grain and load three more, each time monitoring for indicators of excessive pressure. When pressure is encountered I back off one full grain and consider that charge to be maximum. When I've completed this I move on to another bullet and repeat the whole process, continuing until I've tested all combinations with one powder. Then I switch to another powder and repeat the whole process.

At the conclusion of all this, which takes a lot of time and considerable shooting, I'm usually confronted with several bullet/powder combinations that produce similar accuracy. Three-shot groups with each of these are fired again to check my initial results, then I sit down and analyze the results. In some instances my best accuracy may come with low or medium powder charges while other combinations are most accurate with near maximum charges.

In the case of my .30–06, which I'd be using on game in open country at relatively long range, I ruled out all but the hottest loads. Accuracy was nearly equal, of course, because I was after a load that will deliver that bullet on target at 300-plus yards with enough remaining velocity to give good penetration and expansion. Then, too, maximum velocity produces the flattest trajectory, minimizing long-range error in bullet placement due to error in range estimation.

At this point your choice of loads is whittled down. In my .30–06 I had just two that I felt deserved more testing. You'll notice that I've not mentioned the use of a chronograph to measure the velocity of my loads. There's a simple explanation for this. While it's very nice to know what the muzzle velocity of a load is, it's neither essential nor pertinent to load performance. You can get some idea of what your loads are doing for velocity by comparing them with similar loads in the reloading manuals. If you have access to a chronograph, by all means use it, but if such facilities are unavailable, don't worry about it.

Now that I've settled on two loads for further testing, I head back for the reloading room where I can work in comfort and guaranteed precision. I load up 15 rounds of each load, then go back to the range and sight-in with each load so the bullets hit approximately where I want them to at 100 yards. Next I shoot three-shot groups with each at 100, 200, and 300 yards, allowing the barrel to cool completely between the firing of each group. The reason for testing the loads at these distances is because the load that groups best at 100 yards isn't always best at longer ranges.

As I shoot these groups, I pay careful attention to where the first shot hits because this is the shot from a cold barrel and tells me what to expect on

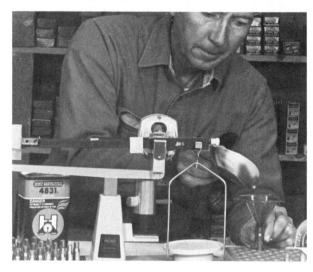

After deciding on two loads for final tests, Milek moves back into his shop for more development.

Milek uses measure to dump bulk of charge into scale pan, then trickles the last few tenths of a grain to "zero" the beam. Consistent charges are readily duplicated in this manner.

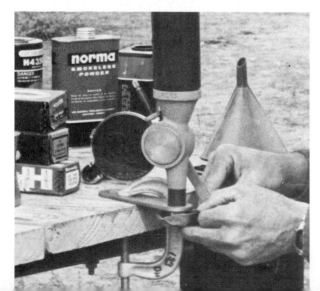

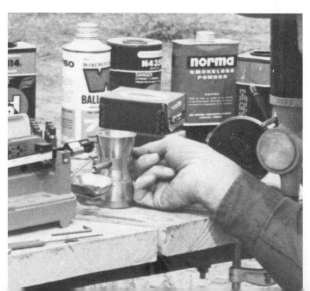

To maintain accuracy of data collected in this series of tests, author scrubbed the bore thoroughly after each three-shot group. A fouling shot was then fired before continuing with the next string. The hours of toil are really a labor of love, manifested in fine groups (right), which were fired at 100, 200, and 300 yards. The flyer in target's lower right corner was called by the author when fired.

a shot at game. After all of the groups are fired, I study the targets carefully and finally choose the load I'll hunt with. In the case of the .30–06 load for hunting mule deer and elk with the T/CR rifle, my choice was a charge of 59.0 grains of H4831 powder behind the 165-grain Nosler solid base spitzer bullet.

One more step is required before load development is complete. Sight the rifle in at the distance you choose, then shoot groups at other distances to see where your shots will hit. With my chosen load in the T/CR .30–06, I found that if I sighted it to hit 1 inch high at 200 yards my shots would hit 2½ inches above point of aim at 100 yards and 5 inches low at 300 yards. Sure, you can use ballistic tables to gather this data, but the results won't be as accurate as those obtained through actual shooting.

As you can see, working up a load for big-game hunting isn't difficult, but it does take a lot of time. There are a few points that still need discussion, though. The first involves cleaning the bore during firing tests. If you don't clean the bore, fouling from both powder and jacket metal will soon have an adverse effect on accuracy, casting doubts on the accuracy of your work. I clean my bore after each three-shot group, first scrubbing it with a brush dipped in Marksmans Choice, then following with a couple of patches soaked in the same solvent. I follow these with dry patches, then one soaked in Outer's Crud-Cutter, a fast-evaporating solvent that leaves the bore dry and free of grease and oil. Finally, before firing for group, I put a fouling shot through the barrel.

Second, what do you strive for in the way of accuracy in a hunting load? The best you can get, of course, but don't throw up your hands in disgust if your three-shot groups don't measure an inch or less at 100 yards. Too many big-game

hunters have been led to believe that nothing over an inch group is acceptable. I don't buy that. To begin with, few light sporters will shoot this well. Then, too, no hunter can take advantage of such accuracy from field shooting positions. However, I do feel that you should insist your rifle produce groups under two inches and if it doesn't, have a gunsmith do some tuning work on it.

Then there's the matter of which load to choose; the most accurate or the one giving the best velocity? Here a lot depends on shooting conditions. If your shots at game will come at close range where maximum velocity and a flat trajectory aren't of much concern, then by all means choose the most accurate load. But for long-range shooting you want all of the velocity you can get, so I'm usually willing to trade ½ inch of accuracy for more velocity. More often than not, I'm not confronted with this decision; I usually find at least one top-velocity load that delivers accuracy on a par with the best lower velocity loads.

Finally, is it essential that you try a variety of bullet makes and styles when developing a hunting handload? Of course not! If you have a favorite bullet, you need to work only with it. The same is true of powder. But remember, the fewer combinations you try, the less your chances of finding the most accurate hunting load.

As any hunter knows, a big-game hunt is comprised of about 99.9 percent walking, waiting, spotting, and stalking and 0.1 percent shooting. Why, then, should you devote hours and hours and hundreds of rounds to the development of a big-game hunting load? It's simple—because unless you do, you greatly increase the chance of all of that walking, waiting, spotting, and stalking being for naught. The best job of hunting in the world is wasted if you can't hit your buck when the chance finally comes!

Reloading Steel Shot

Don Zutz

The age of Steel Shot has been relatively slow in dawning, but things are moving faster now. More and more zones are being changed to steel shot only, and some states—such as Nebraska—have either opted for steel shot entirely on waterfowl or are leaning in that direction. One state (Wyoming) has a bill pending which would require steel shot for all hunting, upland as well as waterfowl! And although steel shot still has an element of controversy about it, the trend seems irreversible. There is a move to steel shot that may well snowball during the next three to five years. It is something shotgunners will have to accept if they wish to continue hunting.

One of the biggest raps against steel shot has been price. Factory-loaded rounds with steel pellets have become very dear, and, to make matters worse, it has been virtually impossible to reload steel until recently because of a lack of information, advanced concepts and components, and poor distribution of whatever steel reloading information and equipment did exist.

But all of that has been changing with a rush. Breakthroughs have been made, and before long any hunter will be able to reload steel shotshells as readily as he now reloads with lead shot. It will not be a matter of using the same wads and powders and recipes once employed with lead shot, of course. Steel shot introduces new needs because the pellets do not compress like lead. The wads and loads, meaning powder selections and charge weights, must be suited specifically to steel pellets. It is an involved story, but one that we will attempt to unravel here. To begin, let's hasten to add that if you intend to throw charges of steel shot from the charging bar and bushing assemblies of press-type equipment, you'll need some accessories to expedite matters. Whenever we reload with bulky shot sizes like BBs, 1s, 2s, and 4s, even lead pellets do not flow very fluidly from the hopper into the charging bar and down the drop tube into the shotcup. The first problem comes with charging-bar movement, which often means binding and shearing with bulky pellets that do not permit easy bar movement. This condition worsens with steel pellets, because steel doesn't compress or shear like lead shot under pressure. If steel pellets jam the hopper neck or the bushing, movement is sometimes impossible. And the tendency for large pellets to bridge in the drop tube is another agony! Sometimes these bridges can be broken loose by rapping the drop tube, but sometimes they remain tight and necessitate the use of a stick to work them loose. Then, when the bridged pellets do free, they have the nasty habit of dribbling all over the floor!

To improve pellet flow from the hopper to the charging bar and then slickly through the drop tube requires greater dimensions at the narrow points. At least one shotshell-press manufacturer is already on the market with such widened accessories for steel shot. This is Mayville Engineering Company, Inc., maker of the popular MEC line of presses. I assume other press makers will follow shortly with conversion kits of their own. Since only the MEC kit is handy at the time of this writing, I will base my comments on it. The reader must simply realize that, should other press makers follow suit, the overall concept will be the same, namely, to expedite smooth pellet flow with large-diameter, noncompressible pellets.

The people at MEC actually have two elements relative to steel shot. One is a kit to convert the press assembly for improved pellet flow, while the second is a set of charging bars made solely for steel pellets. I'll start with a review of the press conversion kit.

This article first appeared in Shotgun Sports

Steel-shot reloader is shown pouring Ecoshot into reservoir. This is "soft" steel shot—as it should be. Steel pellets should have a rating of no more than 90 DPH (Diamond Pyramid Hardness). John Sill photo

MEC produces a kit with necessary parts for converting a reloading press of MEC origin from lead to steel shot. Part of the conversion kit is a new, wider drop tube (left) for smoother passage of steel pellets. Wad ram of steel-shot conversion kit (right) also has been changed. It is wider and cylindrical to eliminate or reduce bridging.

The MEC kit has three components. There is first a new plastic bottle with a larger neck for better flow. The unit no longer uses a grommet to prevent shot shearing, and the increased diameter of the bottle's neck reduces the tendency of pellets to jam between the bottle neck and the charging bar. A second feature is a new drop tube designed with a special shape to mitigate against bridging at the top. Part three is a continuation of the drop tube, being a newly designed wad-ram tube with a straight inside diameter to prevent bridging in the lower segment of the drop tube/wad-ram coupling.

In general, these MEC kits are made to handle steel shot down to BBs. If the pellets get any larger, jamming and wedging can occur. However, if the pellets get any larger, they are normally considered buckshot sizes anyway, and they are then loaded according to count, not volumetric displacement.

Interestingly, the MEC steel-shot conversion kit isn't only good for steel shot. I find that it works well with the larger sizes of lead and copperplated lead shot, too, and would heartily recommend it to those reloaders who, through the years, have been struggling with blockages whenever they reloaded pellets larger than 7½ birdshot.

MEC has smartly developed special charging bars for steel-shot reloading. The shot bushings designed for lead shot wouldn't work anyway with steel. Because steel pellets of any given size are lighter than those of the same-sized lead pellets, there will invariably be more steel pellets per ounce of any given size than there will be lead

ones. As a result, enlarged pellet bushings must be used to accommodate the correct amount of steel shot. By the same token, don't make the mistake of using lead shot with charging bars drilled specifically for steel shot, as there will be a considerably heavier load with lead shot which leads to excess chamber pressures.

From what I have seen thus far, MEC charging bars for steel shot are made black, as opposed to the normal red for lead shot. To date, there are no shot bushings; instead, the shot cavity is drilled directly into the bar. The powder cavity has room for interchangeable powder bushings, of course.

Charging bars for steel shot are also made for certain sizes of shot. This is mandated by the fact that large pellets drop lighter charges than finer shot sizes from the same cavity, the reason being that there is more air space between larger pellets. If we filled a given bushing with No. 4 steel and then filled the same bushing with steel BBs, the No. 4s would invariably weigh more because they pack tighter and get more steel into the space.

MEC has accepted this problem and has made special bars for respective shot sizes. My literature lists six different charging bars thus far: two bars are made for 1⅛ ounces of shot, with one being designated for BBs through No. 3 shot and the second being designated for No. 4 through No. 6 shot. There is a pair of 1-ounce bars with one being made for BBs and No. 2s, whereas the second 1-ounce bar is drilled for No. 3s through No. 6s. A duo of ⅞-ounce bars shows one intended for No. 1s through No. 4s, while a second ⅞-ouncer is for No. 6 steel. Thus, the handloader

Using correct charging bar is crucial, and charging bars are made for specific steel-shot sizes to prevent improper charges. For instance, the MEC bar at left is for 1 ounce of shot in sizes from BB through No. 2. Another bar at right is for 1⅛ ounces of shot in sizes from BB through No. 3. (Right-hand photo by John Sill)

must have his wits about him when going from lead to steel. Using the bushing for lead shot is wrong!

Another important accessory for reloading steel shot is a wad-slitting tool. Steel shot requires tough, high-density wads that are much thicker than those used with compressible lead shot, and all these wads come to the handloader unslit. That doesn't mean you're being encouraged to use them unslit. No way! Wads should always be slit! These wads simply aren't given petals by the molds, and you're instructed to slit them yourself. Most recommendations have it that a wad for long-range shooting should be slit three times; a wad for intermediate to relatively long range should have four slits; and a wad intended for more open patterns and closer ranges should have six slits.

The wad-slitting tool is highly recommended because it makes straight, razorlike, longitudinal cuts—emphasis on the "straight." Some reloaders may try to escape the extra cost of a wad cutter and make the slits with a sharp knife, but that invariably produces crooked lines and uneven shotcup elements which, in turn, can affect performance when the wads strike air resistance upon muzzle exit. Using a special wad cutter will even out the petals for uniform "blossoming" of the shotcup and a perfect getaway by the pellets.

Some reloaders believe that unslit wads give tighter patterns because they hold the pellets together longer, but that's a mistake. Unslit wads can flip after exiting the muzzle, thereby trapping

shot inside to make the wad a veritable slug. The way an unslit wad wiggles into the pellets when it strikes air resistance can also upset pattern development by bumping into shot. Steel pellets tend to pattern tightly as it is, so always slit the wads to let air resistance open them and cause them to fall behind the shot string rather than slamming into it.

Wad-slitting tools with three, four, and six blades are available from U-Load, Inc., P.O. Box 443–177, Eden Prairie, MN 55344 in both 10 and 12 gauges. They are made to function via handle pressure on a press-type reloader: the wooden handle of the cutter is placed under the wad-seating ram and, with the wad positioned below, the handle is lowered to make the cut. Each cutter is guided straight into the upright wad by a pilot-like extension on the tool. The whole procedure is easy. If in doubt, start your steel-shot reloading with a cutter that makes four slits in each wad. Some tremendous patterns come from them, and the flare back from wads so slit seems to be evenly produced.

A handloader-type scale is also recommended, because one will want to know exactly what charges his new steel accessories are dropping. I suggest being terribly concerned with the powder charge, as steel loads are less forgiving than lead ones when high pressures are present. Lead shot will swage down and ram through the forcing cone to provide excess powder gases with some expansion room to avoid a burst; however, steel pellets do not swage down or deform under shot-

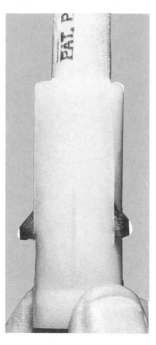

gun pressures, and they are more likely to wedge tight in the chamber or forcing cone when impacted suddenly by extremely high pressures. If the steel pellets won't move, the gases will seek another avenue of escape and could cause a burst in the breech area.

The savvy handloaders's job is using tools, accessories, components, and data which funnel the hard steel shot forward as fluidly as possible.

Such fluid payload movement isn't possible with the wads commonly used with lead shot. These lead-shot wads are molded of soft, low-density plastic and steel pellets under pressure will perforate them like a hot knife goes through butter. The result can be anything from a scratched bore to a load that applies immovable sideways pressure to cause a burst in the breech area. Thus for steel shot, *do not substitute wads designed for lead!* Employ only the long, heavy, deep shotcups designed, molded, and advertised especially for steel-shot reloading. (It's important that you separate lead-shot components from your steel-shot items. Since most handloaders will assemble fewer steel shotshells compared to lead, a reminder list, like pilots use, to check off components should

Steel-shot wads generally come unslit and must be slit with a special tool. Above is the Supersonic four-blade cutter, for intermediate to long-range patterns. Tools are also made with six blades for more-open patterns at close range, or with three blades for tighter patterns at long range. Below: A wad-slitting tool installed in a reloading press quickly can cut straight longitudinal slits with just one stroke of the press handle. (Color photo is by John Sill)

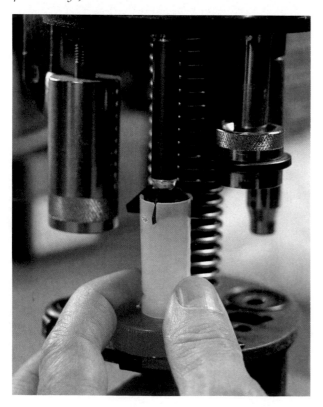

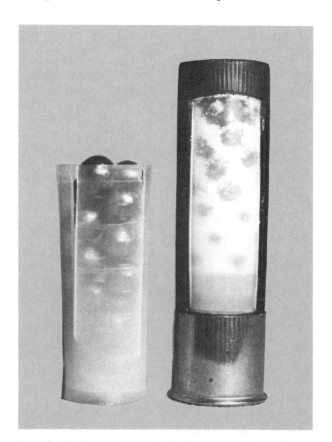

Best hulls for steel-shot reloads are those which are cylindrical inside for optimum capacity. They need to accept the long, deep, and thick shotcup wads made for steel pellets, plus slow-burning powder charges.

guarantee you'll have the proper components on hand before you reload.

Steel-shot wads are available from Ballistic Products, Inc., Dept. SS, Box 488, Long Lake, MN 55356, which makes a tough, deep, high-density steel-shot wad called the TUFF. It is available in 10-gauge Magnum under the title TUFF-BPD and in 12-gauge Magnum as the TUFF-BP12. Both models of the TUFF are vented to help lower the pressure by reducing bore/wad friction. And there's an exciting development coming from the people at U-Load, Inc., Dept. SS, P.O. 433-177, Eden Prairie, MN 55344. U-Load handles the Supersonic® line of steel-shot reloading components, which includes the Supersonic 10- and 12-gauge Magnum wads of considerable toughness and capacity. Prices among the wad makers will be comparable; however, the reader is cautioned against jumping to any conclusions before reading each manufacturer's materials. All sources have instruction books and reloading data for their respective components: hence, contact them for advertising material and pay a few bucks for their instructional materials first. The concepts aren't all alike, and the wads aren't interchangeable on a helter-skelter basis as so many reloaders seem to believe.

It will be noted that steel-shot wads do not have cushioning sections, because the interior space is needed to handle the shot charges. Steel shot of any given size is lighter than lead shot of the same diameter; hence, more wad capacity is needed to host the greatest number of steel pellets needed to achieve a given weight. A 1⅛-ounce charge of steel shot, for instance, commonly requires as much interior space as a 1½-ounce magnum load of lead shot. Insofar as the steel pellets themselves are concerned, the absence of a cushioning section is of no significance, steel shot doesn't deform, so it can take the setback forces. Some cushioning would be good to soak up initial combustion, of course, but the absence of such a factor is worked into the reputable reloading data.

One final point that should be obvious: steel wads *must* encase the entire shot charge! Any steel pellets that lie ahead of the shotcup mouth can scratch the bore. Too, the slits in steel-shot wads must be razorlike to prevent pellets from pushing through to the bore wall.

Powders for steel-shot reloads must perforce be of the slow-burning variety. The theory is based on the need to start the steel pellets moving slowly so that they don't wedge together and become a solid obstacle that exerts sideways pressure to make movement difficult. The transition from chamber to bore via the funnel-like forcing cone is the danger point; if the load of steel shot can be started smoothly and fluidly into the cone, much of the battle has been won. Thus, fast-

Center wad illustrates how lead-shot wads have thin shotcup wall, which would be easily perforated by steel shot under pressure. Wads on either side show how steel-shot wads are made with thicker shotcups.

burning powders can cause steel shot to wedge in the forcing cone or chamber and generate high chamber pressures. There have been reports of hunters blowing up their guns with fast-burning powders and lead-shot wads when they reloaded with steel pellets to "see what it would do." They found out! Soft wads that permit steel-shot perforation under setback forces, and fast-burning powders that give stiff setback, can dynamite a bird gun by putting far too much pressure on steel pellets in the chamber/cone region, and by the soft wads permitting serious perforations.

The best handloader-type propellants for steel-shot reloads today are Du Pont SR-4756, Hercules Blue Dot, Hodgdon HS-7, and Winchester 571 Ball. These are the slowest-burning fuels we can place into a hull while still leaving adequate room for the heavy wads and space-consuming shot loads. These powders should *never* be used in charges recommended for lead shot when steel pellets are employed; powder charges for steel and lead shot are entirely different matters. There can be no interpolations or extrapolations. All powder charges for steel-shot loads must be laboratory-proved solely with steel shot.

Always make certain the powder charge is as recommended in reputable published data. Don't take a powder bushing for granted; many of them throw light and/or lack somewhat in uniformity. Too, some batches of powder will vary in bulk density.

Moreover, do not increase powder charges recklessly in a quest for added velocity. Speed is a necessary factor in steel-shot loads, but steel-shot charges build chamber pressures faster than do lead-shot charges. Lead shot, on the other hand, will compress and swage through a forcing cone and bore even with a somewhat heavier-than-normal powder charge. Steel-shot charges will not swage down due to their hardness. The result is, as mentioned above, a radial expansion of the steel-shot charge to create a virtual blockage and send chamber pressures crashing upward steeply. An excess powder charge that might still shove a lead charge through the bore safely could, therefore, generate burst-level pressures in the breech of a shotgun stuffed with steel. Powder bushings that drop charges which vary by only a few tenths of a grain aren't dangerous, of course, but indiscriminate increments involving full grains can be destructive.

Steel shot should be softer than the gun-barrel steel to avoid damage, a fact that makes shopping for steel pellets a critical matter. As a scientific guide, steel pellets ought to be in the category of "soft" steel with a Diamond Pyramid Hardness rating no harder than 90 DPH. By contrast, lead shot seldom exceeds 30 DPH. However, air-rifle shot checks out at about 150 DPH while ball bearings run about 270 DPH. Neither air-rifle shot nor ball bearings should ever be used for shotshell loads, as their extreme hardness can damage a barrel quickly, especially bulging it at the tight spots such as the forcing cone and choke. I hope that, someday, steel-shot manufacturers and suppliers will stamp the DPH rating on the packaging.

Which hulls for steel-shot reloading? The more spacious types are best, because the wads are big and the shot charges are long. Hulls like the Winchester AA and Peters blue target case, which are very popular with lead-shot reloads, have little application with steel. Far better are tubes like

the Federal Gold Medal, Federal plastic field-style case (Hi-Power), Federal plastic steel-load hull, and even the Federal paper Champion. Also outstanding are cases like the Winchester polyformed, Remington SP, and Winchester or Remington steel-shot hulls. Some of the best 2¾-inch 12-gauge steel-shot reloads I've assembled have utilized the Federal Gold Medal beautifully.

One word of caution about these spacious, field-style hulls: always note whether the reloading data call for a paper-based or plastic-based case. For both Remington/Peters and Federal have been changing the structure of field-style cases, going from the former paper-based case to plastic-based hulls, and the plastic-based tubes tend to run higher pressures for any given reloading recipe. Putting a powder charge designed for a steel-shot load and a paper-based case into a plastic-based tube can create an excess pressure condition. Thus, always eyeball the hull's base wad before reloading with steel, and always read the recipe carefully!

Finally, what size of steel shot should be used? A rule of thumb says that, to get the same *approximate* downrange performance, steel shot must be 2–2½ sizes larger than lead. In other words, No. 2 steel pellets are needed to *approximate* the effect of No. 4 lead. I use the word "approximate" with emphasis because such things as muzzle velocity and range play roles. In my experience, steel 2s falter noticeably beyond 35 to 40 yards, whereas lead 4s can often hit with gusto at 45 to 50 yards provided a hunter can center his bird with lead 4s. Fringe hits by either steel 2s or lead 4s beyond 40 yards tend to mean cripples. As a personal thing, I opt for steel pellets about 2½ sizes, or thereabouts, larger than lead. Over the seasons, I've come to enjoy the No. 1 steel pellet for ducks and geese out to a full 40 yards, and I deem it my basic all-around duck pellet. When there is any doubt about a duck load of steel shot, try 1s!

Number 6 steel shot was not developed for waterfowl, except, perhaps, for a few close shots on decoying birds. Essentially, steel 6s were made for upland or shorebird hunters (rails) who must employ steel in steel-shot zones when not hunting ducks or geese. Steel 6s are quite potent from an Improved-Cylinder upland gun out to 25 yards, but thereafter they slow rapidly and hit like 8s and 9s. Since steel shot is more numerous per ounce than lead, steel 6s can pretty well duplicate the work of hard 7½s. If I have to hunt pheasants or grouse with steel, I'd prefer a steel 6 for the IC chokes and steel 4 for the tighter tube. This is what I used in a steel-shot zone adjacent to a huge waterfowl area, and it seemed to work very well indeed on the ringnecks I took with 1⅛-ounce factory-rolled steel loads.

Wads for steel shot are not like those for lead; they must be deeper and thicker. From left are a Supersonic 10-gauge Magnum wad, Ballistic Products Pattern Driver, NTC 12-gauge unit, and typical Winchester WAA12 lead-shot wad for comparison.

PART SIX

IMPROVEMENTS, ALTERATIONS, AND GUNSMITHING

In Search of an Accurate Barrel

C. E. Harris

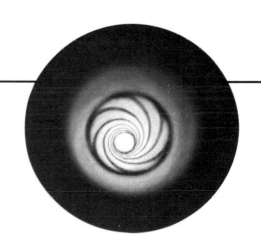

A quality barrel is the heart of an accurate rifle. Good bullets, proper bedding, and precise sighting equipment are of course essential but it is the rifled bore which imparts gyroscopic stability to the projectile. This permits it to fly with its nose pointed almost in the same direction as its trajectory. Without adequate rotation, an elongated bullet would tumble end over end. An accurate barrel must not only provide enough spin but must preserve the balance and concentricity of the bullet in its journey from case to muzzle. Any deformation of the bullet which accentuates its initial yaw upon exit will degrade its accuracy.

There is a great deal of misinformation and folklore about accurate barrels. This is compounded by the general ignorance of shooters as to the processes by which rifle barrels are made. All you need to do to obtain an accurate barrel is visit a major benchrest match, or any national match, and ask what the winners are using. You can buy one from a proven maker of unquestioned reputation, such as Hart, Atkinson (H-S Precision),

McMillan (now Wiseman), Douglas, Shilen, MATCO, SGW, or Obermeyer. Those are the major makers of target barrels used today. There are others, many of whom are virtually unknown, who also produce quality products. Just because a barrel is made by a small maker doesn't automatically mean its quality is second rate.

Following the pack will enable you to buy from an established maker of known quality, but it will not educate you in the nuances which distinguish a "bug-holer" from an ordinary rifled tube.

Among dedicated accuracy buffs there are differences of opinion as to whether a cut-rifled barrel is as potentially accurate as one which is button rifled or hammer forged. Such discussions are popular after dinner at rifle matches, and if you can devote the following to memory you will be

This article first appeared in Rifle

Serious experimenters like Jim Carmichel, shown with a heavy bench gun used for ammunition testing, have done much to improve the accuracy of barrels used today. Most benchrest competition is with rifles of 10 to 13½ pounds, lighter than the monster shown. But the heavy guns are the most accurate.

able to impress your friends and fling your opinions about with the best of them.

There is no inherent superiority of one rifling process over another, *when materials and workmanship* are equal.

If you buy the best, it makes no difference. But an ordinary cut-rifled barrel is better than a poor buttoned one, and a good production buttoned barrel is superior to a poor cut-rifled barrel.

The potential accuracy of stainless-steel barrels is no better than that of alloy steels, when both barrels are blessed with equal physical properties, dimensional uniformity, and bore surface finishes.

A good barrel must start from a bar of steel which, as a minimum, is of aircraft or gun-barrel quality. Use of substandard materials, such as cold drawn steels which have not been heat treated, or material which has not been inspected for flaws in its structure, is an accident waiting to happen. Gun-barrel grade steels have restrictive requirements as to their cleanliness, chemistry, and physical properties. Additional testing and inspection ensures a uniform distribution of sulfides or other additives used to control

A Wilson Arms quality control technician is shown with a borescope, here inspecting the interior finishes of completed barrels. The better instruments used today are fiber-optic units with external power supplies, 10×, having direct or 45 degree oblique lighting. They permit observations of the smallest imperfection, tool mark or flaw in the bore. The very best barrels will look almost perfect to inspection.

machinability, while also ensuring the needed strength and toughness through heat treatment.

A good barrel starts right in the shop where the steel was made. Some mills can produce good clean steel or mirror-finish quality, while a few others simply cannot do the job right.

As for material, suffice to say that most .22 rimfire, shotgun, and low-pressure centerfire barrels (working pressures below 40,000 c.u.p.) are made with carbon steels such as 1137. High-pressure centerfire barrels for sporting use are generally made of alloys similar to 4140 or 4150 chrome-moly, usually in a resulphurized grade. Barrels for military automatic weapons are normally made from a chrome-moly vanadium alloy as defined in military specification Mil.-S-11595. While chrome-moly V steel does not machine particularly well, it is said to be more erosion resistant and is therefore specified for M16 rifle and M60 machine gun barrels.

Most modern target barrels are made of stainless steel such as type 416. Mills catering to the aerospace and nuclear industries are becoming more important sources of specialty stainless steel for the gun trade. Crucible 416R is a proprietary alloy formulated especially for high quality rifle barrels.

Most steel warehousers don't list their products under AISI grades, but use trade symbols which add to the confusion. Names such as B3X, B340, and MAX-EL 3½ all refer to a similar type of re-sulphurized 4140 containing .05 to .15 percent sulphur. Complicating the picture, most steel peddlers know nothing of the gun business and do not know what to sell. Sometimes the folks in the front office writing the order know little more. Most barrel makers aren't big enough to get the attention they should, and have too little pull to get the quality steel they need.

Bigger companies specify exactly what they want and buy an entire heat of 20 tons or more melted and rolled to their requirements. A small shop couldn't even dream of that, not at today's prices. A small maker may take confidence in a brand name, knowing that he is getting good steel which works for him, but he is seldom able to take advantage of newer methods such as argon-oxygen decarburization, vacuum degassing, and ladle refinement, which can be used to control silicates, other nonmetallic and trace elements, while providing optimum machinability, surface finish, heat-treatment suitability, and other desired physical properties.

Small makers usually buy aircraft or gun-barrel grade steel from a warehouser to standard AISI, ASTM, AMS, or military specifications, or by trade names. Bars are already rolled to size, cut to length, heat treated, and stress relieved.

Production shops with their own heat-treating

Major producers of gun barrels for the trade, such as Wilson Arms, buy steel by the "heat lot," color coding the bars to identify the type and batch of steel to facilitate tracking through the process.

department buy bars by the car or truckload, which are then hot-rolled, quenched, tempered, cold drawn to final size, then bumper straightened, magnetic-particle-inspected for flaws, and they expect to do their own stress relief after rifling.

Big shops, which make production barrels for the trade, will take pains to identify steel lots moving through the shop by color coding bars. Having a consistent flow of uniform material is essential, because even steel with the same nominal specifications often drill and ream differently from one batch to the next. That can affect the ultimate size, surface finish, and straightness of the finished barrel.

Relieving internal stresses is essential for barrels rifled by cold forming because those methods set up stresses in the bar which are not present to the same degree in cut-rifled barrels.

Once a steel bar is cut to length and trued on centers to run vibration free on the drilling machine, it is drilled to make the bar a tube. This first step is a trick in itself. The geometry of the drill bit equalizes the forces which would nudge it off center. Ordinarily, the drill remains stationary while the work rotates. Forces are greatest on the outside, so the bit tends to be self-centering, unless hard spots in the bar cause it to wander. On some production machinery dedicated to barrels larger than .50 caliber, dual rotation of the drill and the work is common. This requires use of higher coolant pressures and heat exchanging equipment beyond the means of small shops.

A gun barrel drill is a V-shaped, single-fluted,

Bar stock used for gun barrels comes in mill size bars 20-feet long, which are cut to length in an automatic cut-off machine. Smaller shops buy bars cut to length.

tungsten carbide tool with two cutting edges. The bit is silver-brazed to a hollow oil tube, which is also V-shaped in cross section. The drill is fed about .001 inch per revolution while the work is turning from 1,000 to 2,000 rpm. Cutting oil is forced up the drill shaft through an oil hole in the relief, ground in one facet of the bit, washing the chips back along the V-slot in the shaft.

It is considered normal for a drill to drift as much as .001 inch for each inch of bore length.

Variations in the steel's hardness due to faulty quenching procedures or core segregation make drilling more difficult and cause the bit to wander. When steel quality is uniform, results are obvious.

Drilling leaves a rough bore, which must be smoothed prior to rifling. That is essential, even for cold forming processes such as buttoning or rotary forging which work the bore's entire surface, because microscopic layers of metal are smeared over the bore's surface, and these microscopic "file edges" have little heat capacity and are therefore burned away by the hot powder gases and flake off, exposing a rough microstructure beneath. That leaves a bore which fouls incessantly and has very short accuracy life.

I have observed this phenomenon in some hammered barrels which shot great for a few thousand rounds, then with no warning went from the 10-ring to the target frame in the same 20-shot string at 600 yards! No amount of cleaning could restore the barrel's accuracy.

Drilling is usually .005 to .006 inch smaller than the desired bore size, which of course is determined by reaming. Bore finish prior to rifling is often enhanced in premium-grade target barrels by lapping, honing, or electropolishing.

Bore reamers are the long fluted type which may be either pulled or pushed. Push reaming is favored for producing better surface finishes, but is most practical in calibers smaller than .25, since the rod has a tendency to buckle in larger bores.

Most shooters imagine a barrel shop as being a small crowded building with only a few machines. A production shop, such as Wilson Arms, shown here, is a sizable operation with dozens of drills, reamers and rifling machines.

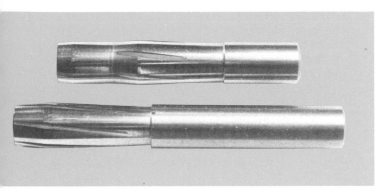

Above are two common types of rifling buttons—a push type on top, and a pull type. Push buttons are separate from the drive rod, whereas pull buttons are silver-brazed to the push rod. Below is a six-spindle conversion of a Pratt & Whitney ½B reaming machine at Wilson Arms Company.

Here a circa-1930 Pratt & Whitney sine bar rifler has been converted for button rifing. This is the best way to obtain uniform twist in a buttoned barrel. The sine bar riflers are also used for cut rifling pressure and velocity test barrels and for experimental work.

Pull reaming is favored by makers who button-rifle. It provides more uniform dimensions which eases the problem of obtaining the right relationship between bore size and button diameter. This encourages a uniform button driving force and reduces the chances of leaving loose spots in the bore.

A bore reamer is affixed to a hollow oil tube which rotates through the bore at low rpm. The reamer is turned while the work is held stationary on a movable table powered by a leade screw (on older machines) or hydraulically.

Some makers ream twice for a better finish, but one pass is adequate when done properly. The success of single reaming depends on the quality of the cutter.

Getting a good finish and maintaining uniform dimensions when bore reaming are absolutely essential if a barrel is to be accurate. Top makers of quality target barrels earn their money right there. Higher priced barrels are honed and/or lead lapped prior to rifling. Some hammer-forged production barrels are both honed and electropolished to obtain surface finishes in the 4- to 8-micro-inch range even before they are hammered!

Up to this point all barrel blanks are prepared

similarly for rifling. The actual rifling processes and handling of the blank vary according to method.

Cut rifling is the oldest method and is still widely used. Early rifling machines used a hand-carved wooden leade screw in which the helix was determined by coiling nitrated rope around a wooden mandrel, then burning it away to aid subsequent handwork. By the early 1800s, wooden mandrels gave way to steel leade screw machines.

Most cut-rifling machinery used today is of the Pratt & Whitney ½B hydraulic series developed in the late 1930s and widely used during World War II. Earlier P&W sine bar machines, dating from World War I and produced into the 1940s, are still in use today. They are highly prized for test barrel and experimental work, due to their great versatility. They are ideal for small shops because they are smaller and lighter (about a ton) than the average three-ton ½B twin spindle hydraulic machine.

In a typical hook-cutting machine, the cutter is contained in a rifling head attached to a hollow oil tube. A steady flow of oil under pressure to the head cools the cutter and aids chip removal, which might otherwise cause galling. Depth of the cut is controlled by a wedge and setscrew which governs protrusion of the cutter from the head. The rifling head is drawn lengthwise through the bore and is indexed after each pass so it will make an identical cut on the next groove, continuing for an entire revolution of the barrel. Cutter depth is then reset, either manually (on older machines) or automatically, and the sequence is begun again.

Each pass of the cutter removes only .0001 to .0003 inch of steel. During the rifling process, the barrel must be indexed and the cutter reset many times. It takes 20 or 30 passes to cut each groove to its proper depth; consequently, hook cutting is very slow.

During World War II, considerable effort was exerted to increase production without having to build large numbers of machines. Today's most common rifling methods, button rifling, broach cutting, and rotary hammer forging all trace their origins to wartime pressures.

The broaching method cuts all the grooves at once by passing a long, progressive stepped cutter through the bore. The barrel is held stationary while the broach, several feet long, is pulled through. Each set of cutters on the broaching rod resembles a short cylinder with flutes cut in it. Each succeeding cutter makes a somewhat deeper cut until bore and grooves are correctly dimensioned.

Broaching is widely used for revolver barrels, and for some military rifle barrels. The biggest drawback to broaching is the cost of tooling. A single .38 revolver broach costs about $1,000, is brittle as glass, and requires a skilled grinding department to maintain it.

Although broached barrels have a reputation for being rough, this needn't be the case. The key is having quality steel of the correct composition and heat treatment to obtain optimum hardness for good finish.

M1 and M14 National Match barrels were usually rifled by the broach-cutting method.

The rifling process developed during World War II which had the most impact on the postwar sporting firearms industry was button rifling—a cold forming process in which a carbide tool, with the rifling ground in reverse upon it, is forced through the bore.

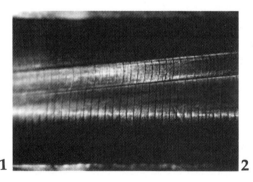

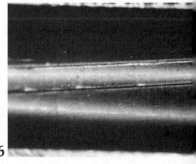

Characteristic circumferential reamer marks are visible in (1) this popular-priced buttoned barrel. Cut-rifled barrel (2) is marred by draw scratches parallel to the rifling. The very best buttoned barrels (3) reveal no visible imperfections, even under magnification. The best cut-rifled barrels are almost, but not quite, this smooth after lapping.

Heckler & Koch polygon barrels (4) are extremely smooth inside. Their accuracy is as good as—but seldom superior to—other styles of rifling of equal quality. Mass-produced barrels such as Remington (5) and Winchester (6) are hammer-forged and boast an interior finish equal to many custom barrels.

The metal displaced as the grooves are formed is squeezed into flutes ground into the button, and becomes the lands.

The original button-rifling process was developed by Remington and was used for some 03A3 Springfield and Garand barrels by the war's end. On earlier tools, two working surfaces were used on the button. The first plowed the grooves. A second, called the "trailer," burnished the tops of the lands. That method often left a burr turned on the edge of the lands and was soon refined into two variations currently employed.

The push method of button rifling, originally developed by Clyde Hart and later improved by G.R. Douglas, features a pilot in the leading portion of the button which aligns the trailer with the rifling helix ground upon it. The pilot may also size and burnish the reamed bore very slightly.

The pull-button method, perfected by the Danjon Company of Cheshire, Connecticut, and used by such makers as McMillan, Shilen, and Wilson Arms, calls for the button to be silver-brazed to the pull rod. In premium barrels, the pitch of rifling is controlled by sine-bar gearing or by a leade screw, in addition to the helix ground on the rifling button. This is the best way to ensure uniform twist rate.

Push buttoning depends more on the helix of the button, since the latter is not attached to the push rod. When button rifling, it is difficult to control twist much closer than ±½ inch without a gear-controlled or sine-bar-driven machine. Cut rifling, however, can provide exact pitch of rifling within ±¼ inch or less.

Ultimate bore and groove dimensions of a button-rifled blank are affected by the diameter of the blank, due to the extreme radial stresses involved. Material hardness, reamed bore diameter, and tooling dimensions must be closely controlled. At Douglas, for instance, a slice is cut off each bar for hardness checks and those which fall outside the desired range are sent back to the furnace for another heat treatment. Buttons are kept in graduated sizes to permit adjustments based on test runs of each heat of steel. The samples must be rifled, air gauged, then distinctively marked so they can be identified after stress relief and checked again for comparison with previous dimensions.

It is normal for a .30 caliber buttoned blank to shrink about .0005 inch after stress relief. Hardness of the blanks must also be monitored so button driving force will be uniform from blank to blank. Given proper process and material controls, button rifling is practical even in .50 caliber machine gun barrels and 40mm grenade launcher tubes.

Proper stress relief is essential in a buttoned barrel to ensure the bore will not "trumpet" or widen toward the muzzle after it is turned down. For that reason, button-rifled blanks are given a low-temperature heat treatment after rifling in a controlled atmosphere to reduce scaling. Some target barrels may be stress relieved after they are straightened, which eliminates their tendency to walk or regain memory as they heat in firing. Better gunsmiths lap a buttoned barrel after profiling to ensure uniform bore dimensions.

Rotary forging of gun barrels was first done experimentally in Germany in the 1930s. The technique was not widely exploited until the late 1950s when modern cold forging equipment was developed for sale on the world market by Ge-

Air gaging is used to measure bore and groove size. Changes in air flow around a carbide plug detect variations as small as .0001 inch. Air gaging is done on a sampling basis to aid tool selection and also to monitor the effectiveness of stress relief.

sellschaft für Fertigungstechnik und Maschinenbau (GFM), of Steyr, Austria.

In this process, the barrel blank is drilled larger than the finished bore size (usually 10mm for a 5.56mm barrel and 11.5mm for a 7.62mm barrel). The blank is reduced in diameter and elongated as it is rotated and hammered around a carbide mandrel which has the rifling, and often the throat and chamber, ground upon it. In shotgun barrels the chamber, forcing cone, and choke can be formed all at once.

The mandrel is inserted through the breech end of the drilled bore and pushed to the muzzle. As the steel is hammered down around it, it is withdrawn slowly until it emerges from the breech. Behind it is a finished barrel, requiring minimal exterior machining. Process time is about 2½ minutes for a typical rifle or shotgun barrel and about six minutes for a 20mm cannon tube. Rotary forging is widely used at Remington, Winchester-USRA, Sturm, Ruger (shotgun barrels), and in Europe by such makers as Steyr, Heckler & Koch, and Fabrique Nationale.

Hammer-forged barrels have been shown to have superior strength under abusive field conditions, such as firing obstructed barrels. This is primarily due to the granular structure of the barrel's steel, which tends to toughen under the severe jolting it takes when hammered. It is aided also by the cleaner steels, essential to successful use of this process, and to the virtual absence of tool marks which can sometimes initiate cracks or other flaws.

When the exteriors of hammer-forged barrels are machined, they react differently than those of cut or buttoned barrels. Because the metal has been worked so hard, the bores of hammered barrels tend to constrict slightly as the barrels' surfaces are machined. This gives a slight taper-bore effect of about .0015 to .002 inch from breech to muzzle of a 24-inch barrel. Many feel that aids accuracy.

I have never been able to prove that. Taper boring is also claimed to increase velocity slightly but chronograph tests of taper-bored barrels' velocities compared to those of standard barrels revealed differences less than the velocity standard deviation of the ammunition.

Claims for higher velocity and improved barrel life for some hammered barrels are based on reduced gas loss around the bullet, which is aided by having a slight draft or slope on the sides of the land. That leaves the top of the land narrower than its base chord. Its form eases production during the hammer process and also improves obturation of the bullet and reduces its deformation by eliminating sharp corners in the bore. It reduces fouling, too.

An advantage claimed for button-rifled and hammer-forged barrels is that cold working results in a smoother bore finish than can be produced by cutting. In the best barrels, that is true. The bore of a Hart or McMillan target barrel has a mirror finish. The finish of HK polygon bores is comparably superb.

In inexpensive buttoned barrels, the circumferential tool marks caused by reaming are not entirely removed by passage of the button. Because the marks are at right angles to the direction of bullet travel, they are more harmful in terms of fouling and bullet deformation than the longitudinal scrape marks of a cut-rifled barrel.

A typical military broach-cut barrel has an interior finish of about 32 micro-inches. This will gradually improve if the barrel is regularly cleaned and occasionally scrubbed with mild abrasive paste such as Brobst JB, Flitz, or Simichrome metal polish. The accuracy of such a barrel will improve steadily for the first several thousand rounds when properly cared for.

Honing or spin-polishing a reamed bore will reduce surface finish to about 16 micro-inches, the same as the finish of most pressure/velocity and accuracy test barrels used in the industry. Lapping such a barrel after rifling will further reduce surface finish to 10 or 8 micro-inches which is equal to that found on button-rifled barrels such as Shilen and Douglas. Special procedures such as the double lapping Obermeyer uses in his 5R barrels, or electropolishing, can help approach surface finishes in the 6- to 4-micro-inch range. Zero finishes as in a set of gauge blocks or polished lenses are undesirable, as they greatly increase friction due to the extreme surface contact. A barrel can be *too* smooth.

The process of barrel lapping is an area burdened by more than its share of old wives' tales; it is nothing more than a simple polishing process. Most people who get poor results try to use too fine an abrasive. The most successful uses of lapping are to provide uniform dimensions, rather than to improve finish. If it's done well, you get the finish as a side benefit. Lapping can help a rough barrel, and it is useful as a cleaning technique to salvage barrels which are heavily fouled, rough, or neglected. It will not, however, make a good barrel out of a bad one. In fact, if done improperly, the process can ruin a perfectly good barrel. It can make an acceptable barrel from one which would otherwise be poor and can often improve one which is already good. There is no point in using anything finer than AA Clover (about 240 grit); a perfectly adequate lapping job can be done with B or A grit, depending on the type of steel and its hardness.

A rough barrel can shoot well if properly and regularly cleaned. A smoother barrel may not group any better, but will maintain a consistent bore condition longer between cleanings and will give better accuracy life because there are fewer

COMPARISON OF CUT RIFLING, BUTTON RIFLING AND ROTARY FORGING
(+) or (−) to indicate advantage or disadvantage

Cut Process

+ Variety of bore and twist variations possible at modest cost

+ Can turn before rifling to reduce risk of warping or dimensional changes

+ Near-perfect concentricity of bore and groove possible

+ Twist is exact and repeatable from barrel to barrel

+ Cost competitive in small lots, well suited for experimental and prototype work, adaptable to small shops

− Rifling machinery complex and not easily improvised. Hook cutting machines out of production for 25 years, parts must be made.

− Process slow, limits production, is costly for large runs.

Button Process

+ Great cost savings on mass production of like units

+ Buttoning machinery simple and can be homemade

+ Buttons readily available in trade channels, and those of simple type are easy to make

− Cannot turn before rifling

− Concentricity of bore a problem with less than four grooves

− Twist tends to be more variable from barrel to barrel and within any given barrel

− Button-driving force and requirements for tooling variations for specified bore and groove limits, practicality of short runs of special items.

Hammer Process

+ Best concentricity of bore to OD of barrel

+ Best potential surface finish possible with adequate blank preparation

+ Fast process. Forms chamber, rifles bore and rough contours barrel all in one operation

− Machinery and tooling very expensive, precludes small shop use

− Different size blank and process information needed for each caliber and contour change

− Requires special expertise in metallurgy and heat treatment to obtain adequate stress relief and suitable bar stock as starting material.

surface imperfections to be cut and eroded by the hot powder gases. My experience does not clearly favor one type of rifling over another but rougher barrels, like those found in ordinary production rifles, usually require more conscientious cleaning to maintain target accuracy.

While stainless-steel barrels are more resistant to corrosion, they will pit if ammonia dope or Blue Goop is left in the bore too long, or if subjected to salt water, chlorate primers, or black-powder residue without proper cleaning. I have seen stainless barrels damaged from galvanic reaction by leaving them heavily fouled for as long as a year in a polluted urban atmosphere.

The advantage of stainless steel is primarily that the types developed for rifle barrels boast a mirror finish and will give the smoothest bore regardless of the rifling process used. Stainless barrels are more easily damaged by lackadaisical cleaning methods despite their rust-resistent qualities.

Whether stainless really gives longer accuracy life than chrome-moly in moderate calibers such as the 5.56mm, .308 Winchester, or .30-06 is open to argument. In higher intensity calibers, stainless might offer an advantage but it would probably be slight. An accurate barrel must be smooth; its bore dimensions must be uniform and the steel should have been properly stress-relieved. Whether it is cut, buttoned, or hammered, or whether it is of stainless or chrome-moly, is of secondary importance.

There is a new process for rifling barrels called Electrochemical Rifling (ECR). In this process the barrel is drilled and reamed in the usual way, and the grooves are formed by electrochemical action. The barrel blank serves as the anode and is connected to the positive pole of a rectifier. The rifling tool is the cathode and is connected to the negative pole. The space between the bore and the tool flows with an electrolytic solution which is forced through the bore at high velocity. The bore surface facing uninsulated parts of the tool is chemically dissolved away to form the grooves. This process is being investigated by Remington and some other makers, and has attraction in mass manufacture, owing to its fast cycle time of about 25 seconds and the fact that the process sets up no additional stresses in the blank beyond those incurred in drilling and reaming.

A disadvantage of the process is that interior surface finish is limited by the quality of the bore reaming and by the electrochemical process. A rough bore ream will leave tool marks across the tops of the lands. The rifling is not sharp and the grooves seem slightly rough, appearing in a borescope to have been sandblasted minutely, with a surface finish of about 25 micro-inches. I have experimented briefly with one of these barrels on a 14-pound .308 bench gun, and it would not group better than minute of angle. The same lot of 168-grain Match ammunition fired in ordinary production buttoned barrels chambered with the same reamer would shoot half-minute or better. And the best benchrest-grade barrels would stay in the high .2s and mid-.3s! I think this process has yet to be proved and will probably see more use in production of low-cost arms than for high priced custom or target rifles.

Alvin Linden, Dean of Stockmakers

Ludwig Olson

Alvin Linden died 40 years ago, but he is still remembered for the high quality of his stocks. He is also remembered as a designer and artist who set high standards and gave inspiration to members of the stockmaking clan. His well-written booklets on stock inletting, shaping, and finishing told very effectively how to do those tasks, and recorded his methods and designs for posterity. Unfortunately, he died before his manuscript for the booklet on "checking" was completed. (He used the term checking instead of checkering. More on that later).

Linden was of humble origin. He was born June 28, 1886, in Motala, Östergötland Province, Sweden. His father was a skilled cabinetmaker and boatbuilder, which undoubtedly had a strong influence on Alvin's career. According to Linden, "I graduated from a public school in Sweden, and completed a three-year course of manual training under the Naas system. My instructor was a strict taskmaster who insisted on perfection, so I learned the woodworking trade very well.

"I was taught how to square a piece split from a block, and to work from two right-angle surfaces, just as in precision patternmaking and gunstocking. All this was done by hand. Finally, I made cabinets with blind dovetail joints in drawers. I also built boats."

As to his liking for guns, he recalled, "My father never had guns, and didn't like them, so I was given strict orders never to have anything to do with them. Being bull-headed, it was only natural that I'd do just the opposite. While still very young, I got a hatchet and pocketknife for Christmas. My father told me how to chop my fingers, not how to avoid it. So I didn't chop my fingers, and have not done so since."

Linden's father emigrated to America in 1892. In August 1899, Alvin, his mother, a brother, and two sisters came to Pullman, Illinois, where the

This article first appeared in Rifle

Linden was famous for his unorthodox method of roughing out a stock. All he depended on was a razor-edged hatchet, a keen eye—and a lifetime's worth of experience. (Dave LeGate illustration)

father was employed by the Pullman Palace Car Company. During 1899–1900, Linden attended school in Pullman. He went with his father to Newark, Ohio, in 1900 to build a new railroad shop. In 1901, he returned to Pullman where he spent almost five years in the Pullman shops working on interiors of sleeping, dining, and private cars.

Many kinds of wood were used in Pullman cars. Linden recalled having worked with kokko, white mahogany, red mahogany, several types of walnut, oak, satinwood, rosewood, and tonquin.

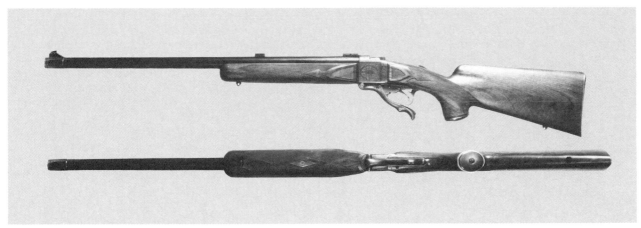

Linden-stocked Farquharson

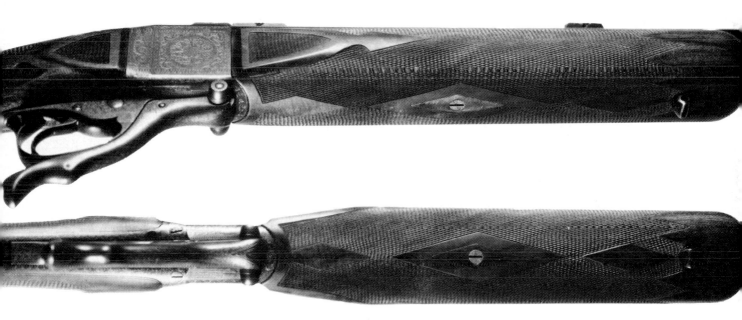

Details of a Linden-stocked Farquharson

Most difficult of all to work was tonquin, a very hard, dark green wood filled with fine sand. It was grown in Southeast Asia, but Linden and his fellow workers claimed it must have originated in hell.

The inspector at Pullman was a cranky old Scot who had especially sensitive fingers. If the work didn't suit him, it had to be done over again. Producing work of high quality to satisfy the meticulous inspector wasn't easy. But evidently Linden did very well. While still in his teens, he was one of the workers on the prize dining car exhibited at the St. Louis World's Fair in 1904. The experience at Pullman proved very valuable in his later career as a stockmaker.

About 1907, Linden had an accident that caused him to become partially crippled. He backed into the stub of a limb while sawing wood, and it injured his spine. This later developed into a rheumatic condition, which caused difficulty in walking and greatly impaired use of his left arm. Evidently this accident occurred after he had moved to northern Wisconsin, where he worked many winters as a sawyer and saw filer in logging camps. For three years he was a saw filer at Neopit on the Menominee Indian Reservation, where he earned the reputation of being the best in the business.

Although a born gunbug, he didn't start making stocks until 1920 or so. At that time, and until his death, he lived with his sister and brother-in-law, Donald J. (Dan) Grant, on a small farm near

Bryant, Wisconsin, about 11 miles northeast of Antigo. Linden's small frame shop, covered with tarpaper, was behind the log house. The place was far from fancy, and one had to travel a narrow dirt road to get there. Nevertheless, it was destined to become a mecca for gun lovers, especially for those with a weakness for fine, beautifully stocked sporting rifles.

Linden (leaning against car) and Emil Koshollek, photographed beside Linden's shop in January 1943.

A meticulous person with the desire to see "a place for everything and everything in its place," would have been horrified by the cluttered interior of Linden's shop. There were heaps of tools on the workbench, but Linden could find whatever he wanted in an instant. He made many of his own tools, which often proved better than those purchased. Some of his chisels, for example, were made from worn-out files. He also made special purpose tools such as scoring hooks. Most of his stocks were produced solely with hand tools. A few years before his death, he acquired a band saw and drill press to perform some of the less critical work.

One might wonder how Linden, with his severe physical handicap, managed to make fine stocks. He had a very strong will. Most of his strength seemed to have concentrated in his right arm. His powerful handshake and steady, penetrating gaze exuded confidence.

Linden was far from satisfied with his first stocks, but his work improved, both in workmanship and style, with each new attempt. One of his early stocks for a 1903 Springfield had a curved buttplate taken from a Winchester Model 1876 rifle. He later graduated to improved, more sophisticated designs developed principally by Ludwig Wundhammer, a Los Angeles gunsmith, and Colonel Townsend Whelen.

In 1929, W. Herbert Dunton, an artist from the West, wanted a rifle built for his daughter. Linden was selected for the job at the suggestion of Colonel Whelen. Dunton had his own ideas about the cheekpiece design of this Springfield sporter. Unlike the usual cheekpiece that curved sharply up-

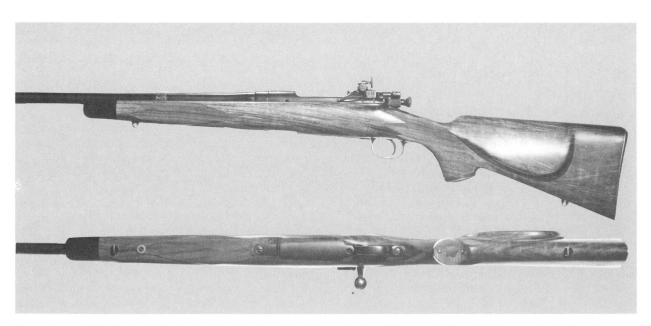

Two views of 03 Springfield stocked by Linden. Simple, uncluttered lines and fine detail were his hallmarks.

ward at the front to meet the top of the comb, the Dunton cheekpiece flowed forward gracefully to blend into the stock at the pistol grip. According to Dunton, this cheekpiece was harmonious in mass and line, and did not disturb his sensibilities. It was generally similar to the cheekpiece on schuetzen rifles and early German sporting guns. Linden offered some resistance to Dunton's design at first. Later, however, he favored it, and the Dunton cheekpiece became a standard feature of Linden's stocks unless the customer desired otherwise.

Ulrich Vosmek, an old master gunmaker of Bohemian extraction, also had an influence on the design and production of Linden's stocks. Vosmek lived in Antigo where it was convenient for Linden to visit. Linden learned a lot during those calls, especially about stock finishes.

The lines of Linden's stocks embodied conservatism combined with elegance and good taste. He had an eye for beauty, and claimed that anyone who could appreciate the lines of a Petty girl (sexy paintings of curvacious girls by an artist named Petty, whose work was very popular in those days) was fully qualified to lay out and model any curves required on a stock. A feature he particularly despised was the sagging "shadbelly" bottom line of schuetzen buttstocks. There is no doubt he would have disliked the flat-bottomed forearms and exaggerated, baroque lines of many sporting rifle stocks produced today.

As to buttstock profile, he insisted that the bottom line should run straight from buttplate toe to a few degrees about the latitude of the trigger-guard tang. It was his contention that the bottom line could either make or spoil a buttstock, and swinging this line down to the lowest level of the grip cap would result in an awkward, paddlelike appearance.

The top of the buttstock was also a straight line from buttplate to comb, unless the customer wanted a Monte Carlo. Drop at heel and comb was fashioned according to the customer's desire as long as it was within reason. Linden disliked excessive drop and thin, razorlike combs typical of many German sporters. He preferred a fairly thick bomb with fluting to provide a comfortable place for the thumb. The pistol grip was moderately full, with about 3½ inches from its toe to the center of the trigger. It was generally fitted with a steel cap. A Wundhammer swell, a bulge on one side of the grip to provide comfort for the hand, was featured on some Linden stocks.

Slightly concave in the middle, the buttplate on the typical Linden stock was well rounded at the top. It was usually thin steel, finely checkered. Linden liked the custom steel buttplates imported from Germany, but preferred those produced by master gunsmith Emil Koshollek, of Stevens Point, Wisconsin. He also preferred Koshollek's checkered steel grip caps.

Linden disliked a skinny forearm. Most of his stocks featured a forearm of moderate width and depth, with the bottom line in a straight taper toward the front. The forearm was rounded on the bottom. It was large enough for the hand, but was not clublike or clumsy. Some of his stocks had a schnabel fore-end, but most were fitted with a rounded hard rubber or Bakelite tip. Ebony and buffalo horn were also used for the tip, but horn was not favored because of its tendency to shrink and crack.

Linden stocks were made from mahogany, maple, cherry, and myrtle, but his all-time favorite was walnut, especially well-figured French walnut. He was very particular about selecting stock blanks, and insisted that they have fancy figure only in the butt, with a straight grain through grip and forearm. It was very important, he claimed, that the grain run slightly upward toward the front of the stock. This gave strength through the grip, and minimized warping of the

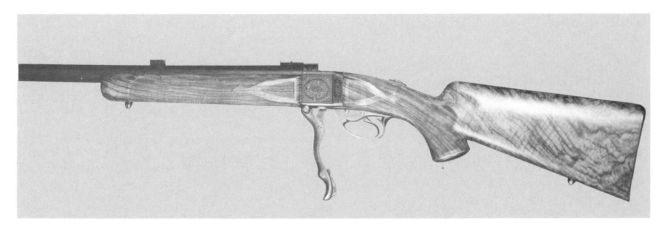

Gibbs-Farquharson single-shot stocked by Linden with extra-fancy walnut.

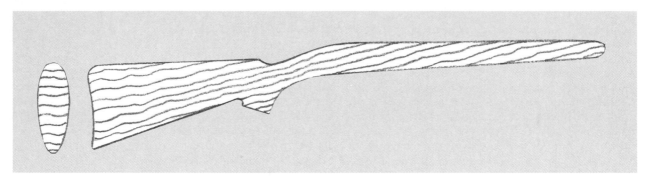

Linden insisted wood grain should run slightly uphill to add strength at grip and toe while minimizing forearm warpage. The cross-sectional view shows grain pattern of quarter-sawed wood.

forearm. It also reduced any tendency to split behind the receiver tang or break off at the butt-stock toe.

In stocks of American military rifles, the wood grain was almost always parallel to the barrel. This was, and still is, also true of stocks on many American factory and custom sporters. Linden called this "musket layout," and condemned it in no uncertain terms.

He also rejected the opinion, held by some, that only air-dried blanks are suitable for making stocks. It was his contention that both air drying and kiln drying are necessary. He pointed out that an air-dried blank has the most moisture in its center, and the moisture will not work outward as long as there is damp air around it. In one of his booklets he mentioned that a blank should be air seasoned, then kiln-dried until it contains seven percent moisture by laboratory section tests, not surface tests. He was considerably more scientific and progressive in his thinking than most stock-makers, and refused to follow tradition blindly, unless doing so offered advantages.

Another important point in Linden's choice of stock blanks was to select quarter-sawed wood with the grain at right angles to the buttplate. A forearm with this grain tends to warp vertically. This is far better than lateral warping, which causes the forearm to put pressure on the side of the barrel.

Inletting on a Linden stock was the ultimate in precision woodworking. Before starting, the right side of the blank was planed flat and true, and the top was planed to square up with the right side. This was followed by sawing excess wood from the bottom back to the grip, and also squaring this surface with the right side. A centerline was then scribed around the top, bottom, and ends of the blank. Accurate measurements for the inletting were made from this centerline.

Principal tools used to inlet the blank were a scriber, marking gauge, scoring hook, router, back saw, brace and bit, chisels, gouges, scrapers, mal-

let, try square, carpenter's square, clamps and a pocketknife. The scoring hook, a curved tool with a sharp, chisellike end, was used wherever possible to make deep cuts at the edges of the inletting, especially for the barrel channel. Most of the rough inletting was then done with chisels and gouges between the scored cuts. Final fitting was done chiefly with scrapers.

Unlike many gunsmiths, Linden made very little use of rasps because wood fibers, after rasping, have a tendency to spring upward if the wood becomes wet. He used sections of old rifle barrels, with edges filed sharp, to scrape barrel channels, and found this method very efficient. Prussian blue and kerosene, or lampblack and kerosene, was used to detect high spots where wood had to be removed.

When asked how tightly the wood should fit the metal parts, Linden would reply in his earthy lumberjack manner, "Tight as the rear of a steer in fly time!" It is true that the wood to metal fit of his stocks was extremely close in most areas, but not throughout. The sides of the magazine box, for example, cleared the wood by a small amount. There were very slight clearances at the rear of the receiver tang and on the ends and bottom of aperture rear sight bases. Linden found that a generous clearance behind the receiver tang, as in many military and sporting rifle stocks, was unnecessary if the ends of the magazine box were fitted tightly to support the wood behind the recoil lug.

Linden was not an advocate of the full-floating barrel system so often used in hunting and target rifles. On most of his stocks he used a screw to secure the forearm to a lug on a barrel band. The upper edges of the forearm were a snug fit with the barrel for appearance and to keep out dirt and rain. There was also a snug fit in the region of the barrel band. A small clearance, however, was present in other areas between forearm and barrel.

There are three principal bearing areas with the

Linden system of bedding for Mauser-type bolt-action rifles: bottom of the receiver tang, bottom of the receiver just behind the recoil lug, and bottom of the barrel a few inches behind the forearm tip. This system works well if the wood is quarter-sawed, properly seasoned, and kiln-dried to control the moisture content, and the grain runs slightly upward in the forearm. Other areas of the bedding must, of course, be correct.

The outside of the stock was cut to shape with a turning saw (very narrow-blade saw with wooden frame), draw knife, and planes. If the wood was not too burly, Linden cut most surplus wood from the forearm with a sharp hand axe. Watching this being done was enough to make a stunt man wring his hands with anxiety! Linden was not one to waste time by removing surplus wood like a mouse nibbling away delicately on a piece of cheese.

As in inletting, Linden very seldom used rasps to shape the exterior of a stock. An exception was when other tools failed to perform as desired on hard curly maple. After removing surplus wood, scrapers and sandpaper were used to achieve the final size and desired smoothness. The stock was then wetted with water, dried to raise the grain, and the "whiskers" removed with fine steel wool. This was repeated several times until the grain would no longer raise. Sandpaper was not used in this process because it tended to press the whiskers down instead of cutting them off.

One of Linden's pet hobbies was experimenting with stock finishes. As in many other matters, he had a scientific approach to the subject, and did not follow custom blindly. Until his booklet on finishing stocks was published in 1941, most gunsmiths considered that rubbing many coats of linseed oil into the wood with the palm of the hand was the only proper method. Some custom gunmakers even held this time-honored method to be sacred. Old traditions die hard, and there are still those who believe that rubbing in linseed oil with the hand is the only way to go.

Linden pointed out in his writings that a linseed oil finish is not waterproof, as often believed. In addition, it has a tendency to become sticky, particularly in hot and humid weather.

Many alibis were offered by those who failed to get the desired finish with linseed oil. According to Linden, a linseed oil finish is like the liniment prescribed by a veterinarian to cure the spavin on a horse's leg. When the horse owner complained that the liniment failed to cure the spavin, the vet would reply that the liniment hadn't been rubbed in enough. If, on the other hand, the horse owner said that he rubbed the spavin a lot, the blame was put on having rubbed too much. And there were other alibis, such as having used too much, or too little, liniment.

A special method was used by Linden to obtain a fine finish with linseed oil. He applied and rubbed in the oil with a woolen cloth because the cloth was naturally oily, and did not absorb the oil. After several coats of raw linseed oil were applied to the stock, with a week's time to soak in, a thin coat of boiled linseed oil was rubbed on and allowed to dry for 24 hours. This was followed by several more coats of boiled oil, with necessary drying periods between applications, until the stock appeared to be lightly varnished. Very fine abrasive paper was then used with water to give a satin-smooth finish.

After experimenting with many different finishes, Linden found that a combination of linseed oil and spar varnish proved best. He applied several coats of linseed oil to the stock, with drying periods between applications. Then, spar varnish was wiped on in very thin coats with a cloth, allowing each coat to dry. Each coat was removed almost to the wood with fine abrasive paper and water. The resultant finish was considerably more durable, more water resistant, and less sticky than a straight linseed oil finish.

Linden showed exceptional skill in checking stocks. As mentioned near the beginning of this article, he used the term checking, instead of checkering. He insisted that square, or checkerboard, spacing was checkering, while elongated, diamond-shaped spacing, used commonly on stocks, was checking. He claimed no special artistic talent was required to learn how to check, or checker. But it was a decided asset, he said, to be born with the patience of Job, possess the asininity of a stubborn mule and have a minimum of common sense. Also, he claimed it was an advantage to blow off steam fluently without conscious effort when things went wrong. The famous Lou Smith of Ithaca Gun Company said that about one person in 12 has what it takes to be good at checking. Linden pointed out that some women, due to their light touch and good control, are apt to be better at checking than many men.

Since most surfaces on stocks are rounded, Linden recommended the beginner should practice checking a birch or maple rolling pin. His method of providing a flexible straightedge for guide lines was to use glued paper tape, or masking tape, applied to the rounded surface in spiral fashion.

He made some checking tools by grinding three-corner files to the required shape. A small V-shaped parting chisel, with the edges ground to the required angle was used to point-up the checking in corners and other difficult areas. He stressed the importance of using tools with good steel, tempered properly to hold a sharp edge. But he said it was no longer necessary to quench red-hot steel in urine before drawing the temper, or to temper blades by running them into bodies

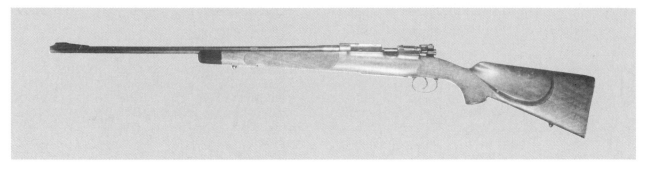

Three views of a Mauser .270 Linden stocked for the late gun and hunting writer Jack O'Connor. Note the Dunton cheekpiece and Koshollek grip cap.

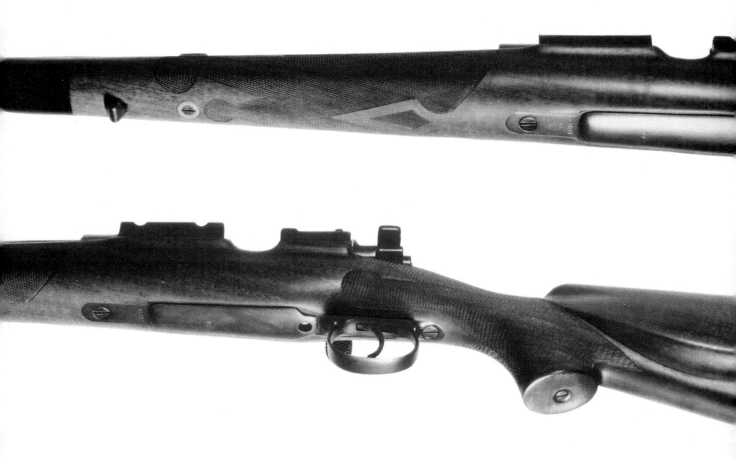

of live slaves, as was reputed to have been done in ancient times by Vikings and possibly others.

Teeth were cut on the bottoms of Linden's homemade checking spacers and V-shaped plows with a metal checkering file. He found that checking tools with handles made from soft pine or balsa enabled him to check longer without a numb feeling settling in his fingers.

There was, and still is, a notion among diehards that checking hides the grain. Linden rejected this notion. He claimed that checking, done properly, with fully pointed diamonds, then oiled, actually brings out more grain than unchecked wood. He applied common motor oil, thinned with kerosene, to the checked areas. Linseed oil was not used on the checking because it turned gummy and tended to hide the grain.

Linden's specialty was borderless checking, with diamonds fully pointed to the edges of the pattern. This required great care and maximum effort, but lent a very attractive appearance. His two keen-eyed nieces inspected the finished jobs, looking for runovers and other imperfections. They proved very meticulous and efficient.

His checking covered extensive areas of the grip and forearm. Spacing of lines was dependent on the nature of the wood and desires of customers, but it was often 20 to 24 lines per inch. Checking patterns were generally simple, with angles at the ends. On some stocks, however, the pattern edges were curved, and occasionally a fleur-de-lis pattern was used.

The Mauser 98 was Linden's favorite bolt-action because of its sound design and the fact that it did not cause as much difficulty for the stockmaker as several other actions. He particularly disliked the peculiar angle of the rear guard screw in Springfield, Krag, and Enfield Model 1917 rifles because it caused a screw hole alignment problem during inletting. Another feature he disliked was the irregular shape of the Springfield receiver on the left side, around the cutoff. He reshaped this area to a streamlined contour, which did not find favor with some Springfield rifle fanciers.

I first met Linden in 1939, between hitches in the Regular Army. Just prior to that time I visited gunsmith Emil Koshollek in Stevens Point, and he suggested we drive to Linden's place. That was the beginning of many visits and a close friendship with two of America's best gunsmiths. Unexpectedly the association with Linden had an influence on my military career.

In early 1940, Linden started work on his stock-making booklets. The publisher was Thomas G. Samworth, then in North Carolina. Samworth was the leading gun book publisher of that time and had been Chief Editor of *The American Rifleman* for a few years during the 1920s. In 1916 he was an army officer on the Mexican border. He was an old-timer in the gun field well known for his direct, impulsive, and peppery manner, along with plenty of wit and charm. His letters contained many expletives, but were always very informative and entertaining. Linden's letters to Samworth were masterpieces in their own right, packed with sound information and laced with crude lumberjack humor. These men met only through correspondence, but they combined their efforts to produce the world's finest literature on the art of stockmaking.

About the time Linden started work on his first booklet, I became bored with duty at Fortress Monroe, Virginia, and requested transfer to the Philippines. But Linden had hoped I would remain in this country so that it would be convenient for me to make drawings for his booklets. Samworth intervened by making arrangements with an old army buddy, Lt. Col. (later Maj. Gen.) Julian S. Hatcher, to have me transferred to Ab-

erdeen Proving Ground, Maryland. Lt. Col. Hatcher, then Commandant of the Ordnance School, assigned me to the job of making charts and maps. In September 1940, I transferred to the newly formed 23rd Ordnance Company at Fort Sheridan, Illinois, where there was better opportunity for promotion.

I wanted to make drawings for Linden's booklets, but there was little opportunity to do so. Just finding a suitable place to do such work was difficult because I was often on maneuvers, living in tents. While at Fort Sheridan I had the job of post Range Sergeant in addition to regular company duties. Moreover, I spent most of my free time with a girlfriend, and finally had thoughts of getting married. When I informed Linden of this, he replied immediately, "Don't do it! Being hitched in double harness might be just wonderful for very short periods of time, but otherwise it's pure hell from what I've observed."

It finally developed that Linden made his own drawings. He had great artistic talent with pen and ink, as was obvious from the many fine illustrations he produced. His work was so good, he finally acquired the nickname "Ole Skratch," or "The Old Kratch." He seemed upset because I didn't make the drawings. This soon wore off, however, after I explained my situation and told him he had done a much better job than I was capable of doing.

After Linden died, his brother-in-law, Dan Grant, was going to complete the manuscript for the booklet on checking, but the job was never accomplished. Dan gave me a copy of the unfinished manuscript along with a well-checkered forearm Linden made for a Farquharson single-shot rifle—reputedly the last stock job Linden worked on. These are, of course, treasured souvenirs.

The three completed Linden stockmaking booklets have been out of print for many years. After Samworth's publishing business was taken over by Stackpole Books, Linden's three booklets were consolidated into one volume called *Restocking a Rifle*. That fine book is also out of print, but copies might be available from gun book dealers.

In 1941 Linden restocked a Winchester Model 70 in 7×57 Mauser for me. The stock blank, a well-figured piece of French walnut, was purchased from Howard Clark of Stevens Point. The checkered steel buttplate and grip cap, each with a trap, were made by Emil Koshollek. A top scope mount, designed and produced by Koshollek, was fastened to the left side of the receiver and to the rear sight notch on the barrel lump. That rifle was the finest I ever owned.

Linden was kindhearted to a fault, and generously provided detailed information on stockmaking to others, even to competitors. He also loaned out rifles to neighboring farmers during deer season. He made a mistake when he loaned a very fine restocked Enfield .30-06 to a neighbor, though. It was in perfect condition when it left Linden's shop. When it was returned a few weeks later, the bore was badly rusted and pitted. Linden said nothing to the neglectful neighbor, but vowed such an incident would never be repeated.

He restocked a German 13mm antitank rifle with the sporting stock made in proportion to the monstrous 37-pound tank killer. It was hung on the wall of his shop as a curio. When someone wanted to borrow a rifle, Linden would tell the party that the monster on the wall was available.

One of the rifles I acquired in 1940 was a deluxe .30-06 Oberndorf Mauser sporter in perfect condition, except that a small piece of wood was chipped out behind the receiver tang. Another fault was that the buttstock was too long for me. It was such a beautiful rifle, I sent it to Linden just so he could see it. After two weeks, the rifle was returned with the buttstock shortened to the length I wanted. Also, the stock was repaired by gluing a piece of walnut behind the tang, and checking that area to match the checking on the grip. The job was done so expertly that the repair couldn't be detected. I offered to pay Linden, but he wouldn't accept anything. It was a good example of how he would do almost anything for a good friend.

Although Linden had very limited formal education, he was an avid reader, and could talk intelligently on almost any subject. His extensive knowledge, as shown in the many letters I received from him, was truly amazing. He gave me a lot of good advice, almost as though he was my father. One valuable piece of advice was that I should write a book on Mauser rifles. This proved to be a successful project. I can't say his views on marriage were entirely sound, but it's a pity I didn't heed his excellent judgement in several other matters!

After acquiring much favorable publicity as an ace stockmaker, he acquired a number of disciples. One was Charles Golueke, who worked as a newspaper engraver in Green Bay, Wisconsin, and made fine stocks in his spare time. Another was Leonard Mews, who lived only a hundred yards from my home in Appleton, Wisconsin. He was a custom stocker for Weatherby, and then went on his own to become one of the leaders of the field. There were other disciples, or those who were influenced by Linden's high standards to produce superquality stocks.

Linden would readily admit that some of his contemporaries, Thomas Shelhamer in particular, turned out stocks fully equal in quality to his own. He would also admit that some of his designs and methods were used by others, and did not necessarily originate with him. He gained the lion's share of the publicity, principally because of his articles and booklets.

Gun lovers worldwide lost one of their best friends when Linden died of an apparent heart attack on July 4, 1946. He was only 60 and had never married. Rifles with his stocks are now prized collector items. His many followers of the stockmaking clan are carrying on the tradition of high quality he established many years ago. Nonetheless, the mystique of the old master remains. It can be said that he was to gun stocks what Antonio Stradivari was to violins.

Despite his bad back, Linden was a crack shot with both rifle and handgun. That two-handed hold was considered radical in 1943 when photo was taken.

.45 Auto Innovations for Accuracy

Jack Mitchell

The military brass are retiring the .45 ACP 1911 pistol in favor of the 16-shot 9mm Beretta Model 92SB-F, but the .45 remains the overwhelming first choice in combat competition circles across the country. When dressed up with a few new parts, treated to a bit of accurizing by a competent pistolsmith, and fed a diet of quality hardball or 68 Hensley & Gibbs ammo, this ageless beauty is transformed into a veritable tack driver. With the growth of combat handgunning as a sport, a number of excellent innovations introduced from the private sector have improved old Slab Side's dependability, reliability, and ac-

curacy. Unfortunately, there also have been any number of gadgets and widgets marketed in attempts to capitalize on the popularity of the 1911 Colt Government auto.

How do you tell the difference between a worthwhile accessory for the Model 1911 auto and a useless, even dangerous gadget? First, determine the primary purpose of owning the gun. A 1911 auto carried for self-protection only remotely resembles a competition handgun. Depending upon the purpose of the gun, talk to the

This article first appeared in Gun World

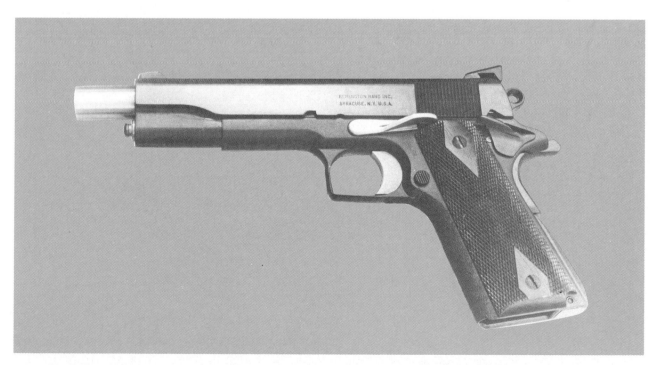

This .45 ACP is equipped with Centaur Quadra-Lok barrel system, and Novak fixed front and rear sights.

pros in either defensive handgun tactics or competition shooting to get their advice. Better yet, see how they have their particular handguns set up. You can bet they have selected particular accessories through an exhaustive process of elimination, choosing those items that work best for specific application.

Notable products developed in the past year for the Model 1911 auto are the Quadra-Lok drop-in match bull barrel manufactured by Centaur Systems of Washington, fixed rear sights for both competition and carry versions by West Virginia pistolsmith Wayne Novak, and a competition magazine well by the same craftsman.

The 1911 auto shown in these pages is an old Colt frame and Remington-Rand slide equipped

QUADRA-LOK SYSTEM

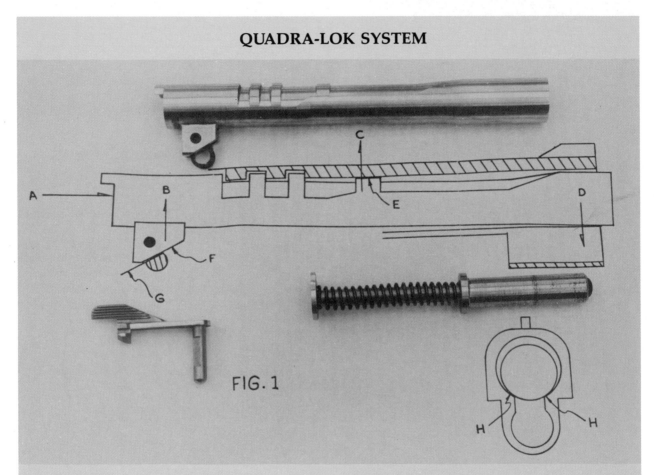

FIG. 1

The Quadra-Lok system is unique in that it replaces the original barrel and barrel bushing, as well as the captive recoil spring and guide and the slide stop. The accompanying diagram illustrates how the unique design works. As the slide moves forward and returns to battery, it contacts the rear of the barrel and pushes it forward in direction (A). This action causes the wedge surface of the underlug (F) to contact the flat ramp surface (G) on the slide stop pin and this action translates some of the forward barrel motion into an upward motion (B) of the rear of the barrel. The special power band (E) contacts the upper inside surface of the slide and presses against this surface in direction (C).

At the instant the power band contacts the slide, further upward motion of the rear of the barrel is translated directly into downward motion (D) of the barrel

muzzle. This downward motion continues until the muzzle contacts the two areas (H) in the slide. When simultaneous contact exists between the muzzle and slide and the power band and the slide, the barrel becomes a stabilized relative to the slide.

The slide will continue its forward travel, causing a continued upward thrust (B) on the barrel. This action continues until the entire slide is lifted high enough to take up the vertical play in the frame rails. The slide then comes to rest with all clearances "jammed" into a rigid position and the forces (A), (B), (C) and (D) remain present under recoil spring pressure. My test gun has worked reliably, using several different types of ammunition and point of impact remained consistent throughout my tests to indicate the Quadra-Lok system works as designed.

Shown in Ransom Rest is Gold Cup .45, factory-accurized version of Colt .45 auto. This one has been further accurized by Jim Clark. On target in front of rest are Clark .45 long slide with Bo-Mar rib, Clark long barrel with fitted bushing, and Dwyer Group Gripper. Pistol grips are by Pachmayr. (John Sill photo)

with Centaur's Quadra-Lok barrel system, as well as Novak's new fixed sight and magazine well. I find the pistol dependable, reliable, and cosmetically appealing. I was quite impressed with consistent five-shot groups offhand at 25 yards resulting in less than two-inch groups. Since the Quadra-Lok barrel I selected has a six-inch length, it lends itself to Mag-Na-Porting which should

improve rapid-fire accuracy and faster recovery between shots.

The Quadra-Lok system is actually a "drop-in" barrel arrangement requiring no fitting. Anyone who can fieldstrip a 1911 auto can install it. Weighing 25 percent more than a standard barrel, it helps to reduce recoil and speeds sight recovery between shots. A self-compensating system pre-

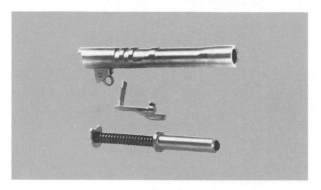

Quadra-Lok barrel system (above) includes barrel, slide stop, recoil guide, and spring. Installation is easily done. Below: Colt frame and Remington-Rand slide are shown before any custom work had been accomplished for accuracy's sake.

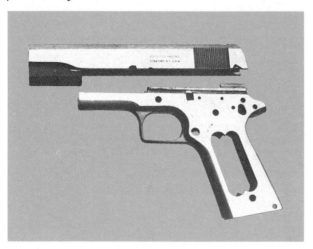

vents the gun from eventually shooting itself loose and the barrel locks up in precisely the same spot relative to the slide after each shot, a key to accuracy in this model.

Three choices of barrel lengths—4½, 5, or 6 inches—offers options that include extended length from the end of the slide to help eliminate recoil, faster recovery time, and Mag-na-porting. A captive recoil system incorporates extra-light, light-medium, and heavy recoil springs to allow custom tuning to particular ammunition. With a retail price ranging from $129 to $139, depending upon barrel length, the 1911 auto owner can save more than double the normal costs of new conventional barrels along with the costs of accurizing by a pistolsmith.

Accuracy with any handgun depends largely upon the type of front and rear sight. Many military veterans armed with the Model 1911 auto complained that they "couldn't hit the broad side of a barn door at 10 paces." Although most service sidearms have a considerable amount of slop built into them purposely so they will function under adverse conditions, the major problem with the 1911 auto always has been the standard military fixed sights. They were designed primarily to be snag proof. They achieved that goal, but are sadly lacking in producing a good sight picture, the critical ingredient for handgun accuracy.

Wayne Novak has designed two versions of his 1911 automatic fixed front and rear sights. The "carry" version of the Novak rear sight is modified from the competition sight, with all edges beveled for nonsnag fast draw during a lethal

Bulk and weight of Quadra-Lok adds 25 percent more weight to muzzle area for faster recovery between shots in rapid-fire exercises. This is an aid to reducing recoil. Inset photo: Flat cut in slide-stop pin aids in lock-up of the Quadra-Lok installation.

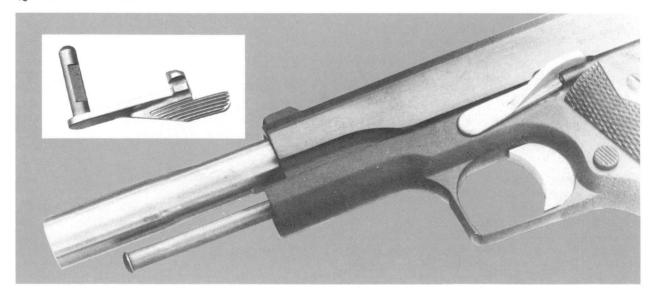

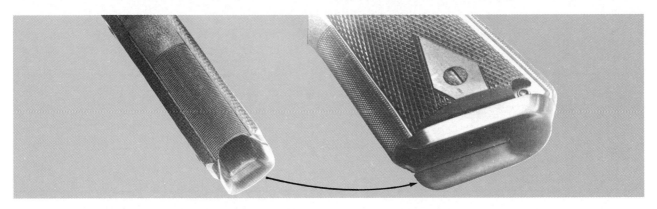

Novak-Smith competition magazine well increases well's aperture by 200 percent. Unit is fitted to frame and silver brazed to mainspring housing for durability. By careful hand-fitting, all gaps and seams between magazine well and frame are eliminated for reliable operation.

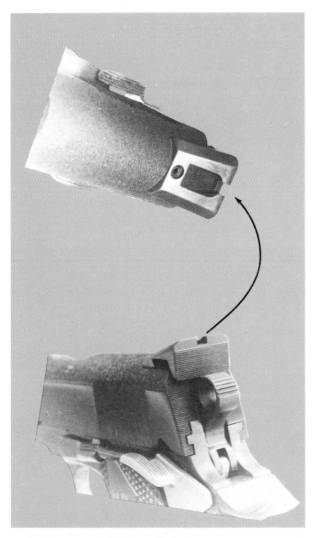

Novak fixed rear sight extends over rear of slide with low profile to offer excellent sight picture. Novak sight fits existing dovetail in slide (top), is anchored by 5–40 lock screw.

confrontation. The Novak rear sight fits the standard 1911's rear sight dovetail and is anchored with a 5-40 locking screw. It can be installed on Model Series 70 and Series 80 Colts as well as the new Colt Officer Model .45 ACP.

Both the carry and competition sights are designed to present a sharp sight picture aided by an extended sighting radius over the rear of the slide. The sights are machined to exact tolerances, hand polished, sandblasted, and finally hot-blued for a matte finish to prevent glare.

The Novak rear sight will not work with the small silver front sight that is standard for the military-type 1911 auto. Novak removes the standard front sight and silver brazes a .200-inch-high steel blank in its place, then contours this replacement for either carry or competition use.

The Novak fixed front and rear sights offer the best sight picture of any I've tried. Ken Hackathorn, noted firearms instructor, calls them the "state of the art in fixed rear sights for carry or competition." Front and rear sights installed and contoured to customer specifications are $85. The competition magazine well designed jointly by Wayne Novak and his assistant, Matt Smith, is a steel funnel device that opens the magazine well for faster reloading. The unit is permanently attached to the mainspring housing and hand-fitted to the individual frame, an extremely strong and durable accessory. The unit may be removed by disassembling the mainspring housing from the 1911 auto frame and replacing it with a standard housing. They are offered in both stainless or blued versions at $95.

Adding accessories that will increase a gun's durability, dependability, and accuracy must be considered in relation to critical safety features. Before adding any to your sidearm, make certain that they meet the above criteria and are of quality materials.

What Ever Happened to Aperture Sights?

Karl Bosselmann

The great prevalence of the open rear sight assembly on commercial firearms is a curious thing. Certainly the gun companies are first to recognize the economics; they are cheaper to manufacture than the apertures. However, one would think that hunters—as well as shooters in general—would attach a highly beneficial peep sight after buying the rifle.

Oddly, most shooters are somewhat ignorant concerning this type of sighting arrangement and tend to make its use complicated and often incorrect. One must not look at or attempt to focus the peep, but rather look completely *through* it. See only the front sight and the game animal or target. Consciously forget about the rear sight; one's eye automatically centers the bead in it. So used, it is a fast and simple system.

Advantages of the close-to-the-eye, quality, receiver-mounted aperture sight over the open styles are listed below:

● Micrometer adjustments are much more precise and more quickly and easily made.

● It is much faster to use.

● Older eyes can use it better and longer.

● It is far more conducive to accuracy.

● Increased sight radius diminishes the "teeter-totter" aiming action and allows for more precise shooting.

● They are much better for running or moving targets.

● In case of damage in the field, a new or spare pretargeted bridge assembly can be easily installed in a brief time. Utilization of an indexing/support screw allows the gun to remain sighted-in after the bridge change.

Many individuals complain about the stuck-on, boxlike appearance of these sights, but such is almost impossible to avoid. The aperture sight is

This article first appeared in Gun World

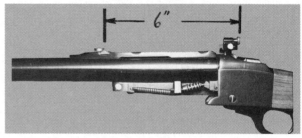

Sight radius may be increased by six inches, more than 30 percent, by addition of Williams FP-98 bridge and rear aperture sight. Open rear sight remains, "just in case." Bridge is right-hand-side mounting.

Two Ruger No. 1s, chambered for powerful .458 Winchester Magnum, display interchangeable front sights. The smaller factory bead, left, measures $\frac{1}{16}$-inch, for smaller game, long range. Custom $\frac{7}{64}$-inch bead for larger, closer, dangerous game. Beads, measured from centers, must each be exactly same height.

an example of the incompatibility of attractive appearance versus extreme practicality.

For example, Winchester fitted a bolt-mounted aperture sight on some of their Model 65 and Model 71 lever-action rifles. It was fine looking, compact, and didn't detract from the flowing lines

This old (pre-War) Model 70 is shown with equally old Lyman 48 aperture sight—a famous combination. Its micrometer adjustments are precise and easy to use, though author feels that protruding knobs can be knocked out of zero. He prefers screwdriver-turned micrometer adjustments. (John Sill photo)

**OPEN
REAR SIGHTS**

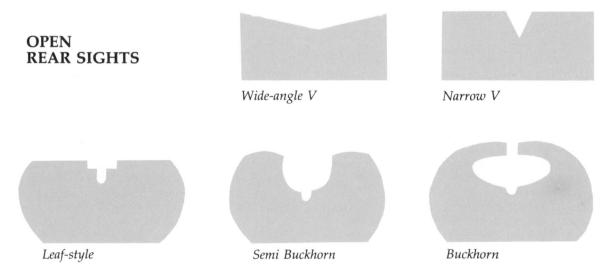

Wide-angle V   Narrow V

Leaf-style   Semi Buckhorn   Buckhorn

The British **wide-angle V** was popularized in Africa and is used in close hunting of dangerous game in dense bush. The **narrow V** rear sight is slow to bring into action because field of view is too restricted. The common **leaf-style** sight features a spring-set folder. Although the **semi buckhorn** sight has drawbacks, the **buckhorn** sight is a disaster because most of the background is blocked out and available light on small notch is minimal.

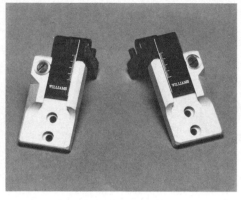

Left: Writer Karl Bosselmann designed prototype rear-sight bases shown above. Base is mounted low on side of receiver wall, ensuring that rear of bridge remains parallel with breech face. Setup permits maximum sighting radius while maintaining adequate access to action and chamber areas. Middle: Here are left- and right-side rear-sight mount assemblies for Ruger No. 1 single-shot rifle. Bridge assemblies are Williams FP-71 on left, and FP-98. Right: The Williams Twilight sight features unusual small brass eyepiece ring that picks up more light during dawn and dusk hours.

of the firearms. But, it was fragile; the weak point being the rear of the dovetail section which would readily bend or break during accidental impact. The sights were soon discontinued, replaced in the marketplace by the more cumbersome boxy receiver-mounted types due to their durability.

As for testimonials for the aperture sights, let's start with our own American legend, the late Elmer Keith, who stated in one of his books, "This is the fastest, most effortless aim possible with any type of iron sight."

Another hunting legend, John "Pondoro" Taylor, opined, "Men out here (and in many other places) have a notion that these are only suitable for target work. But the peep sight is the best and quickest of all iron backsights. . . .

"The peep sight's only real disadvantage is said to be that it cannot be used in bad light; yet I have on more than one occasion killed animals with the aid of a peep sight when it was so dark that I couldn't use the open sights at all. But I always insist on extra-large apertures on all my peep sights."

The minimal requirements for an aperture sight assembly should be as follows. It should be screwdriver-adjustable. Knobs can lose zero by contacting various pieces of gear or by rubbing against arms and legs. It should be mounted as closely as is feasible to the shooting eye without interfering with the accessibility to or functioning of the action. Also, position depends upon the caliber; the sight must not be mounted so far back to be driven into the eye during recoil of a large-caliber gun. The support-screw feature is necessary to provide off-side rigidity against impacts as well as its being an index when replacing the bridge.

It should have lock screws to secure adjust-

ments and an aperture hole of .125 inch should be investigated as the minimum size. It is not advisable to close-fit to the bead. Many individuals, myself included, seldom use the aperture, choosing to remove it and sight through the ring; this enhances the field of view through the sight.

By the term "aperture sights," I refer to those peep sights mounted on the top or sides of the receiver, or on the bolts of a lever gun, not to the tang sights. Besides being uncomfortable to the grip, those tang-mounted sights have the unnerving tendency to drive into the shooter's eye during recoil of big-bore dangerous-game rifles. They do work out well for benchrest firearms such as some of the Schuetzen and Sharps varieties of specialized target rifles.

To achieve best results, the shooter utilizing iron sights must keep both eyes open—the master eye down the sighting plane which does offer some field of view around the front bead, the secondary eye automatically sighting alongside the rifle which offers enhanced and completed field of view. Both eyes being open is especially important when engaging moving targets and applies to rifle, shotgun, and handgun shooting. A perpendicular, flat-faced, gold or brass round front sight bead without edge beveling I have found to be the best. As a general statement, do stay away from white, ivory, and definitely red bead coloring. The bead should be metal—plastic or ivory material breaks far too easily. One who prefers a scope should, nevertheless, have iron sights installed that are targeted for the ammunition used. Thus, should the scope be damaged during a hunt, the rifle is not out of commission. It's wise to have a set of critical spare parts for each different firearm being used. A lot of hunts have been ruined by not preparing for just-in-case.

Three Centuries of Sling Evolution

Pete Dickey

Slings or carry straps are with us today in miles of leather and woven fabric strips which are eventually run through the swivels of rifles of all descriptions. Even a few shotguns and pistols are found equipped today with such accessories.

When the first slings and swivels were used is anybody's guess, but as early as 1672 an Englishman, one Thomas Venn, in *Military Discipline—*

The Complete Gunner, wrote, "He is to have a good Harquebuz, hanging on a Belt, with a swivel."

Venn's belt was probably an over-the-shoulder carrying strap that, with its swivel and an appropriate ring on the gun, was not much different than the rigs used by the U.S. Cavalry up through the nineteenth century. It, and a saddle-mounted boot for the carbine, were used mainly to see to it that horse, arm, and mounted trooper stayed roughly together.

The year 1672 is, however, not the earliest known use of carrying straps, and it is probably not farfetched to suppose that the ancient Chinese smart enough to make a hand-cannon was smart enough to tie a piece of silk or leather to it for a more convenient means of transportation.

Transportation is, to many, the prime justification for the sling, and no one has thought of a

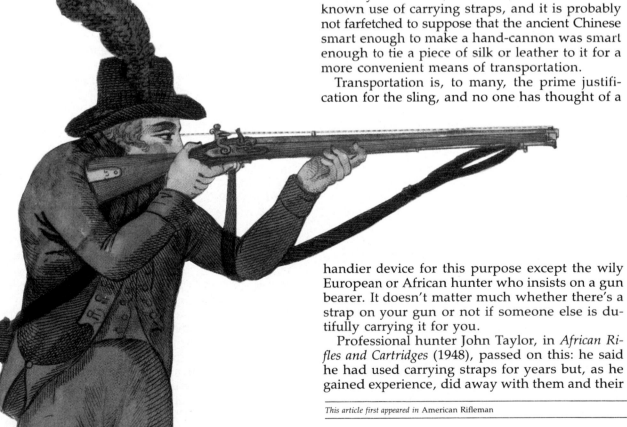

handier device for this purpose except the wily European or African hunter who insists on a gun bearer. It doesn't matter much whether there's a strap on your gun or not if someone else is dutifully carrying it for you.

Professional hunter John Taylor, in *African Rifles and Cartridges* (1948), passed on this: he said he had used carrying straps for years but, as he gained experience, did away with them and their

This article first appeared in American Rifleman

fittings entirely for himself and his bearers. He summed up slings for his style of hunting as "apt to be a nuisance." Many U.S. riflemen, who must struggle through life without gun bearers, would consider the bearer a nuisance and the sling a necessity. Many Continental Europeans, and increasing numbers of Americans as well, feel the same way about slings for shotguns.

These slings—used only for transportation—may as well be called carrying straps. They were called by both names and used by the U.S. military up until the 20th century. U.S. hunters used them, too (Lewis and Clark had 15 gun slings among their supplies), but they were not so common in the field until after the First World War.

Probably the universal use of rifle slings in the war had its effect in accustoming veterans and others to their use. This use, however, was not limited to carrying the rifle. The Army's Model 1907 1¼-inch leather two-piece sling, which has been issued from the nominal year to date, was used before and after the war to steady the rifle in military target practice. Soldiers, sailors, and marines were taught to adjust it so that the loop formed in the forward part, attached to the rifle's front swivel, was tightened on the right-handed shooter's upper left arm so that tension was created between that arm and the front swivel when the rifle was held for firing. The result was a steadier rifle in most positions and, of course, better scores at the range.

The loop feature of the sling was not often used in standing position. Rather the sling, loosened from its taut "parade" adjustment to the "carrying" adjustment, was simply wound around the left arm, giving some extra steadiness. This was and is known as the "hasty sling." In actual combat, except under exceptional circumstances and for *some* snipers in *some* cases, the hasty sling is and was the only nontransportation sling system regularly employed. For this reason, the M1907 sling is issued today only on the M14 rifle intended for snipers. How often its loop is actually employed in sniping is unknown for, as

D. I. Boyd, former Marine Sniper Platoon commander and NRA's assistant director for international competitions, says: "If the sniper has the time, he will probably ignore the sling and 'bed' himself and his rifle into the terrain for as much rifle support as possible."

In *Shots Fired in Anger*, from NRA Sales, competitive shooter and ex-Army sniper John George includes lots of pictures of combat with slings used to carry—but not steady—the rifle in Burma in World War II, and says, simply, "the sling is seldom used in combat." Col. Charles Askins agrees emphatically.

The sling most often issued today is the web M1. It is far lighter and simpler than the leather M1907 (or the M1917 and M1923 web slings that were used to varying extent until the 1942-introduced M1 took over). The M1 is capable of loop adjustment but is not as fine a rig as the M1907 from the target shooter's standpoint.

It is often said that the Americans "invented" the use of the sling as a firing support in the last days of the Krag rifle. This won't hold water, as witnessed by the drawings, reproduced herewith, from gun maker Ezekiel Baker's *Practice and Observations with Rifle Guns*, London, 1804. In describing the plates, Baker said:

"Lying on the belly . . . the sling should be pulled firmly back, to keep the rifle steady. To fire off-hand without a rest, the left hand (should) be forwards on the swell of the stock, the sling under the elbow, which will make it firm and steady." Baker makes no mention of the shooter's hat, which would also seem to be doing its share in steadying the rifle, despite its large target-obscuring plume.

Baker's standing-position use of the sling is no more than a low-mounted "hasty sling." No loop was considered then, and the Americans may well have invented it. Certainly, they have promoted it throughout the 20th century, and target shooters today still insist on it.

Gary Anderson, twice an Olympic Gold Medal winner and director of NRA General Operations, wrote, in *Marksmanship*, that after the rifle and sights, the sling was the most important item of equipment to the target shooter. Forgiving him for treating ammunition so cavalierly, we quote what he said about the sling:

"Many young shooters do not use the sling because it feels awkward, and many other shooters simply do not understand its use. Actually, the sling should be the only thing that keeps the rifle from falling when the shooter is in prone, sitting, and kneeling positions. His left arm is not supposed to make any muscular effort to support the rifle. The sling is a simple leather strap that is attached to the swivel on the fore-end of the rifle and forms a loop that extends around the arm. No expert marksman would think of firing in these three positions without a sling."

To sum up the sling or carrying strap uses and varying opinions, we will offer here three more pertinent quotations.

For the hunter, from Jack O'Connor's *The Big-Game Rifle* (1952):

"A good, well-adjusted sling is indeed handy on a rifle when the shot has to be taken from the sit because of rocks, long grass, topography, etc. The sling is also exceedingly handy when the man is heaving from a long climb, and has to steady down quickly in order to get off a shot. I would not be without one under any circumstance."

For the military man, from *Practical Marksmanship* by Capt. Melvin M. Johnson, USMC (Ret.)—the inventor of the Johnson rifle and machine gun:

"The U.S. Marine Corps has always preached the sling, used the sling on target ranges, won many matches with the sling.

"For combat firing under all conditions you must be able to get hits without the sling for at least one very simple reason: you may not have time to fix yourself in the sling.

"To say that a sling, correctly used, is of no benefit to the shooter's accuracy is absurd. To adopt the proposition that a sling is indispensable for steady, accurate shooting is also ridiculous."

And, for the many of us who have no need of the loop sling but demand a light carrying strap, Elmer Keith, in a letter to Harvey Williams re: the Williams Guide Strap said:

"No laces, no knots, no buckles—and half as heavy. Send me four more."

The dual-purpose "military" loop slings are still available in the original 1¼-inch width for the many competitive shooters and few hunters who still use them. There are also lighter versions of narrower widths, plus simplifications, referred to as "Whelen" slings.

More specialized are the leather cuff-slings that lack the transportation feature and are used only in competitive shooting. Here the design centers around a relatively broad bicep cuff coupled with a short strap that attaches to the target rifle's front swivel. All these target-oriented slings can be found in that equipment bible of the rifle competitor, Al Freeland's catalog.

Hunters' slings are made by many firms, and there is hardly a sizable gunshop around that doesn't carry a large assortment in stock, together with the swivels for those guns not now so-supplied by the factories.

The swivels are of two general types: the "quick-detachables" permit using the carrying strap for transportation and rapidly discarding it for shooting—if that is wanted. More important, perhaps, is that the same strap and swivels can be changed rapidly from one gun to another, provided each gun is equipped with the appropriate studs.

Pachmayr's quick-detachable swivel sets feature studs that are mounted flush with the stock. Pushing the swivel loops in and giving them a half turn frees them for easy sling removal.

QD swivels can be not only convenient but collectible. In 1972, in commemoration of its 25th Anniversary, Michaels of Oregon produced 5,000 cased sets, silver plated and numbered.

Quick-detachables seem to be growing in popularity, but "standard" screw-in swivels, that employ no studs or locking devices, are still common, perhaps because they are cheaper and/or simpler. In order of current popularity, the swivel loops measure 1 inch, 1¼ inches, and ⅞ inch. The two major American producers of both conventional quick-detachables and standards are Williams and Michaels of Oregon.

Another producer of quick-detachables that feature "studs" mounted into, rather than protruding from the stock, is Pachmayr Gun Works.

Bob Brownell, whose catalog is as necessary to gunsmiths and gun tinkerers as Al Freeland's is to competition riflemen, lists and illustrates a good number of carrying straps. Included are Uncle Mike's (Michaels of Oregon) swivels as well as that firm's leather military slings, simple straps, and "cobra-style" straps that, while fitting 1- inch or 1¼-inch swivels, have 2-inch-wide shoulder sections that are suede-faced to resist slipping from the shoulder. Also listed are Uncle Mike's black, brown, or camouflaged wide and narrow strap of nylon webbing, plus a nylon shotgun strap that has self-adjusting loops for barrel and stock and requires no swivels. The famous George Lawrence brand of counterparts of the Uncle Mike's leather slings are also listed, but Brownell proudly "features" his own "Latigo Quick-Set" one-piece strap that can be snapped open for carrying or snapped closed for storage in a matter of seconds.

Williams Gun Sight Co., another prime gunsmith and hobbyist supplier, catalogs its lightweight ⅞-inch Guide Strap in carved, basketweave, or smooth leather, plus a handy two-piece shoulder button that can be instantly applied to any hunting jacket or shirt. It prevents the strap from slipping off even the narrowest of shoulders.

The M1 web sling is, not surprisingly, available

Bob Brownell sells all sorts of slings but, understandably, prefers his own Latigo Quick-Set that is made of one continuous strap that can be snapped open or closed in a matter of seconds.

along with many other military accessories from the commercial Springfield Armory firm.

Reproductions of older U.S. and foreign military slings, including the Civil War carbine strap complete with 2¼-inch buckle, and swiveling snap hook, are offered by Dixie Gun Works, Inc.

Sling'N Things has a nylon or leather arrangement that prescribes carrying the slung rifle or shotgun diagonally across the chest from where it can be rapidly, if unconventionally, shouldered.

The chances are that your favorite holster maker also manufactures slings. The George Lawrence Co. has already been mentioned, but there are many others.

Conventional method of carrying-strap use, with muzzle of the rifle pointed up, is not favored by Williams Gun Sight Co. Here Paul, one of the five Williams brothers, demonstrates use of the Guide strap developed by his father. With practice, a fast transition from transportation mode to strap-supported, aimed firing position is possible and is favored by many.

Left: Al Freeland's experience with target and arm-cuff slings is without equal. Middle: John Bianchi, known better for his holsters, also makes slings. He pioneered the Cobra style. Right: Mag-Na-Port's Larry Kelly sells and uses a sling for serious handgun hunting.

Bianchi makes military slings and simple straps but concentrates on Cobras. John Bianchi registered the name "Cobra" as a trademark in the 1950s. Colors run from yellow through tans and browns to burgundy, finishes from plain through decorative stitching to handcarved, unlined or lined with suede or shearling. Some are made with a cartridge pouch sewn on the face, others have a thumb hole for a convenient grip. Safariland has simple straps and military slings plus padded and quilted "cobras" in full suede (one is green) or smooth leather lined in suede. For the collector/shooter, it also supplies duplicates of U.S. Civil War musket slings.

Old West "cobras" can be had with or without six rifle cartridge loops attached to the sling's face.

De Santis "cobras" are all suede, or suede and shearling lined.

Magnum Sales Ltd. fits swivels, short slings, and Leupold scopes to Mag-Na-Ported .44 Magnum Ruger Super Blackhawks to come up with its Stalker revolver for the big-game handgunner.

The list goes on and new makers and materials will continue to appear, for there will always be target shooters who are truly competitive and hunters who do more slogging than shooting.

The plumed hat may have gone out of style, but not the sling.

Scope That Handgun

Bob Milek

Long and intermediate eye relief scopes for use on handguns of various configurations have done much to bring handgun hunting out of the dark ages. But at the same time, scopes have complicated the life of the handgunner, introducing a variety of problems never experienced by users of scoped rifles. In fact, so involved and unpredictable are the matters of selecting and mounting scopes for handguns, that many shooters haves thrown up their hands in despair and elected to go on shooting with iron sights rather than tackle handgun scope problems.

I've been involved with handgun scopes from the very beginning, and even now I sometimes have apprehensions when I go afield with a scoped handgun. However, the situation today is so much better than it was five or ten years ago that there's just no comparison. In the beginning, scope manufacturers saw the handgun scope market as a small one at best and entered it with inexpensive, poorly constructed scopes that riflemen would be ashamed to put on a cheap .22 rimfire. The result, of course, was scopes that literally disintegrated internally on a handgun generating any recoil. In my gun room, I have a tub full of early handgun scopes that I shot to pieces on big-bore magnum revolvers and specialty pistols.

This article first appeared in Guns & Ammo

Available in four calibers at this time (with more to come in the future), Pachmayr's new Dominator conversion kit for the Colt Gov't Model requires special base/ring mount for scope adaption.

But even though the manufacturers were made aware of the problems, they were hesitant to do anything about them. It seems that no one in the scope business had any idea of the potential for sales that the handgun market posed, so they dragged their feet, hesitant to launch any development projects for fear the investment would be lost.

Today, it's obvious just how underrated the handgun scope market was in those early days. Good handgun scopes now cost as much as good rifle scopes, and shooters buy them without batting an eye. And, thanks to a lot of research and development work, handgun scopes available to the public these days are rugged, reliable optics that, when properly mounted, will give years of satisfactory service.

Handgun scopes are presently available in magnification from $1 \times$, which is actually no magnification whatsoever, to $10 \times$. Nearly all are fixed power, but Burris does offer a $1\frac{1}{2}$-$4 \times$ variable-power model.

The first problem faced by the handgunner who wants to scope his pistol is what magnification to choose. Before a decision can be made, there are a few things that the shooter should know about handgun scopes. First, they have relatively small fields of view, a situation dictated by the long eye relief required. By small field of view, I do mean *small*. To illustrate, consider that most $4 \times$ fixed-power rifle scopes have a field of view approximating 30 feet at 100 yards. However, a $4 \times$ pistol scope has a field of view at 100 yards anywhere from 8 to 11 feet, depending on the manufacturer. Not one scope comes close to having even half as large a field of view as does a rifle scope of similar magnification. Only pistol scopes of $1\frac{1}{2} \times$ have a field of view nearing 30 feet at 100 yards. The smaller the field of view, the more difficult it is to get on target.

As riflemen know, the higher the magnification of a scope, the more difficult it is to use offhand. This isn't because you shake more with a high-powered scope, but simply that your wiggles are magnified and become very unnerving. With pistol scopes, the magnification you can use successfully for offhand shooting becomes even more critical because you wiggle more when shooting a handgun offhand than when shooting a rifle. It's difficult to come up with a hard and fast rule concerning offhand pistol shooting and scope power because every individual has a different degree of wobble to his offhand stance. However, a good rule of thumb is that most handgunners will find it nearly impossible to shoot well offhand using a scope of more than $2\frac{1}{2} \times$ magnification, and a good many find that anything over $1\frac{1}{2} \times$ is nearly impossible. With scopes of $3 \times$ or more, you'll find that a rested position of some kind is needed. When you go above $4 \times$ in handgun scope magnification, a solid rest is nearly always essential.

Then there's the matter of eye relief. Early pistol scopes had what can only be defined as inadequate eye relief for shooting with your arm extended in the normal fashion. In responding to customer complaints about this feature, manufacturers went too far the other way, increasing the eye relief on their pistol scopes so much that shooters with any but the longest arms couldn't get the scope far enough away from their eye to see the entire field of view. Again, scope makers have made some adjustments, but there are still scopes available that won't work on some bolt-action pistols, because they just can't be positioned far enough from the shooter's eye. The high-powered pistol scopes, intended for varmint hunting and benchrest shooting, have what is today termed "intermediate eye relief," meaning that the eye relief is somewhat around 16 inches,

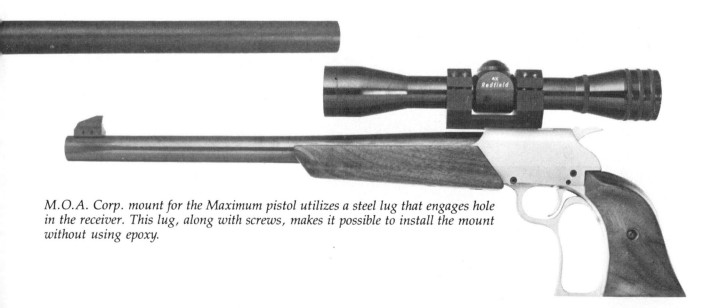

M.O.A. Corp. mount for the Maximum pistol utilizes a steel lug that engages hole in the receiver. This lug, along with screws, makes it possible to install the mount without using epoxy.

rather than the 24-inches-plus common to many lower powered scopes. About all I can suggest where eye relief is concerned is for the shooter to go to a sporting goods store and look at the various scopes available, checking each for eye relief to see that what he chooses will work for him.

Where power is concerned, $1\frac{1}{2}\times$ and $2\times$ are best used for small game at close range and for a variety of big-game hunting, particularly in brushy areas where field of view can be much more important than magnification. Scopes of $3\times$ and $4\times$ are my choice for hunting big game in open country and for small game shooting out to 100 yards or so. Even $5\times$ can be used for such hunting, but I usually set $4\times$ as my maximum for big-game hunting. With their $7\times$ and $10\times$ pistol scopes, Burris stands alone in the market. Granted, these are special purpose scopes, but when it comes to serious varmint shooting at long range, $7\times$ and $10\times$ can't be beat. I recommend the $10\times$ version only to those shooters who've had a lot of experience shooting scoped handguns. Eye placement, both for eye relief and in relation to the optical center of the scope, is extremely critical with $10\times$. Any variation whatsoever from perfect eye placement will result in a total blackout of the field of view.

Naturally, there are no hard and fast rules as to which scope you should use for a specific type of hunting. However, there are some guidelines that many shooters, beginners in particular, will find helpful. The accompanying table lists the pis-

tol scope magnification that I recommend for specific types of hunting.

Mounting a scope on a handgun properly is every bit as important as choosing the right scope—maybe even more so. Today's quality scopes are designed to stand up, so when one fails, the place to look for trouble first is at the mount.

On a heavy recoiling handgun, the forces trying to rip the mount off the gun are terrific. Back in the early days of scoping the Contender, I learned that three 6-48 screws holding the base to the gun weren't enough. A few shots with something like a .30 Herrett or .30-30 would shear three screws off flush with the barrel, and recoil would fling scope and mount right back in my face. The experience was a bit scary, but worse yet was the problem of extracting the broken screws from the barrel so another mount could be attached. Even when Thompson/Center changed from three screws to four, the problem persisted. Screws alone, even when set with Loctite, aren't enough under heavy recoil.

But before we get into how to attach a scope mount to a handgun, let's look at some of the mount systems available and delve into the good and bad features of each. Basically, there are two base systems—rail bases, and those to which the scope attaches with rings as on most high-power rifle mounts. On rail systems, dovetail grooves are cut in each side of the base into which the rail, which is integral with the scope tube, slides. When in place, crossbolts (usually two) are used

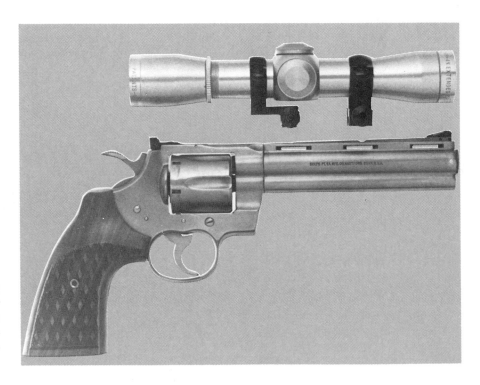

Here's a $4\times$ Leupold scope fit with Colt mount system for the Python revolver. You can see grooves milled in rib to allow proper seating of crossbolts that lock base to rib.

Although the author used a Burris 7× scope on his custom XP-100 to down this pronghorn at 300 yards, he feels the field of view is too small for the novice scope user. He recommends that you use nothing over 4× for big-game hunting.

to squeeze the sides of the base together so they firmly grip the rail. Most often, several shallow grooves are milled crosswise in the bottom of the rail, and these engage the top portions of the crossbolts and help prevent any slippage of the rail in the mount. The use of several grooves provides latitude for positioning the scope in relation to the shooter's eye. Rail-type mount systems are cheaper than base/ring designs, and they're considerably lighter. However, there is one drawback. Even though the dovetails in the base may be milled perfectly square and the rail itself set square on the scope tube, the reticle will often be tipped slightly out of vertical when the crossbolts are tightened. Because the rail is integral with the scope tube, there is no way to straighten the reticle. You just have to live with it.

Base/ring systems are exactly like those riflemen are so familiar with. The base anchors to the gun, then the rings attach by various means to the base. There are, of course, numerous variations to this system. Ruger offers their Redhawk revolver with the base integral with the barrel rib—similar to that used on their M77 rifles—and Ruger rings are supplied with scope model pistols. The rings are clamped onto the gun via crossbolts. Colt offers a mount for their Python, Diamondback, and Mark V revolvers that clamps to the ventilated barrel rib via crossbolts. Two shallow grooves must be cut crosswise in the rib to allow passage of the crossbolts. The grooves also serve to prevent mount movement under

recoil. A most recent innovation, offered by B.F. Products of Grand Island, Nebraska, is a scope base that works on various revolvers and specialty pistols and accepts Ruger pistol scope rings. The base is attached to the rib with four 6-40 screws. The beauty of this one is that the base is low enough that it can be left in place, scope removed, and the open sights can be used. The TSOB mount from SSK Industries, used primarily on Contenders, employs as many as four Bushnell rings to hold the scope to the base.

The bases of most base/ring systems must be attached to the gun with screws. The specialty pistols—Contender, XP-100, Wichita, M-S Safari Arms, etc.—are drilled and tapped at the factory for mount bases. Revolvers are not. Personally, I don't like bases on a revolver that attach to the top strap. It means removing the open sights, then tapping holes in the top strap.

At any rate, if mounting a scope on your handgun requires the drilling and tapping of screw holes, have the job done by a gunsmith. Special equipment is needed to set the base straight and drill the holes to the proper depth. The do-it-yourself tinkerer will usually botch the job.

Now, assuming that your handgun is ready for mount installation, you can handle the rest your-

RECOMMENDED HANDGUN SCOPE POWER FOR HUNTING

Scope Use	1×	1½×	2×	2½×	3×	4×	5×	7×	10×
Small game to 25 yds.	X	X	X	—	—	—	—	—	—
Small game, 25 to 50 yds.	—	—	X	X	X	—	—	—	—
Small game, 50 to 100 yds.	—	—	—	—	X	X	—	—	—
Small varmints to 100 yds.	—	—	—	—	—	X	—	—	—
Small varmints, 100 to 200 yds.	—	—	—	—	—	X	X	X	—
Small varmints over 200 yds.	—	—	—	—	—	—	—	X	X
Big game to 50 yds.	X	X	X	—	—	—	—	—	—
Big game, 50 to 100 yds.	—	—	X	X	X	—	—	—	—
Big game, 100 to 200 yds.	—	—	—	X	X	X	—	—	—
Big game over 200 yds.	—	—	—	—	—	X	—	—	—

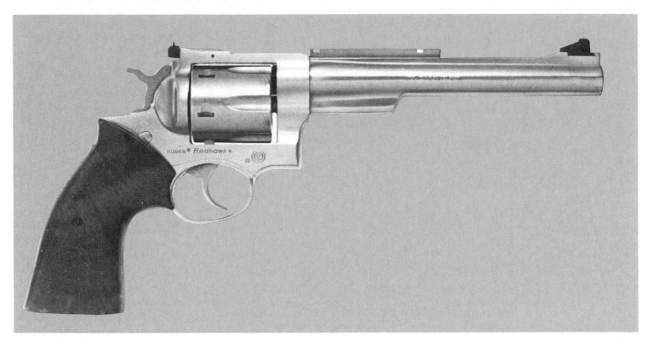

Scope-mount base from BF Products fits several different handguns and allows use of factory sights when the scope is removed. Base uses Ruger pistol rings.

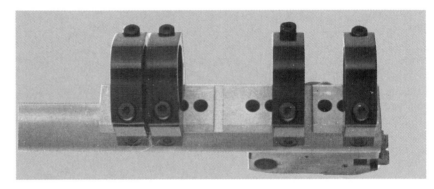

TSOB system from SSK Ind. is anchored to a T/C barrel with six screws and employs up to four Bushnell rings to hold scope.

self. For systems using clamps and crossbolts to anchor to a barrel rib, simply follow the manufacturer's instruction. But on specialty pistols chambered for heavy-recoiling cartridges, additional work is required. The most reliable mounting job is achieved when the mount base is bonded to the barrel or receiver using epoxy glue. Use one of the slow-drying varieties of epoxy because it cures with some resiliency. Five-minute epoxy tends to become brittle and will crack, destroying the metal to metal bond.

Begin by setting the base in place and checking the screws to be certain they're not too long. Tighten a screw and make sure it sucks the base down right. If it doesn't, grind a little off until

the screw no longer bottoms in the hole. Do this for each screw hole, and as screws are sized, lay them aside in order, so you'll be certain to get them back in the holes for which each was sized.

Next, lock the base in place using two screws, and with a sharp-pointed tool scribe the outline of the base on the gun. Remove the base, and with a needle file or emery cloth rough up the surface within the scribed area, removing the blueing. Do the same thing to the bottom of the mount. This creates a surface that the epoxy will bond to readily.

Mix up the epoxy and place beads of it down the center of the gun surface to be covered by the base. You must be careful that no glue gets into

*Mike Wright from Hornady Bullets used a Leupold 2 ×
pistol scope on his Wichita pistol chambered in .30—30
to anchor this fine Wyoming buck. Author feels 2 × is
a good choice of magnification for big-game hunting.*

the screw holes because it will cause the screws
to bottom and not tighten up. Carefully lay the
base in place and tighten the screws. When they're
snugged up, use a light hammer and tap the top
of the screwdriver handle as you further tighten
each screw. This mates the threads perfectly and
the screws will stay tight.

When all of the base screws are tight, soak a
rag in solvent and wipe away any glue that has
seeped out around the base. Be sure to get all of
the glue cleaned off because removing it after it
hardens may cause damage to the surface of the
barrel or to the receiver.

Attach the bottom halves of the rings to the
base as per the manufacturer's instructions. Before placing the scope in the rings, I recommend
that you line the bottom halves of the rings with
something to keep the scope from sliding forward
under recoil. Simply tightening the rings over the
scope won't do the job when heavy recoil is involved. Burris furnishes a sticky piece of paper
for each of their rings, but lining the bottom ring
halves with plastic electrical tape or 400 grit wet/
dry sandpaper works fine. If sandpaper is used,
place the grit side down against the ring.

Put the scope in place, add the top halves of
the rings, and start the screws. Tighten them just
enough so that the scope can still be turned, then
hold the pistol in your hands and position the
scope so that the reticle is square and the eye
relief is perfect for you. Tighten the rings down
securely, but be careful that you don't apply so
much pressure with the screwdriver or hex wrench
that you twist one of those little screws off.

Well, the scope of your choice should now be
properly mounted on your pistol. From here on,
give the rig the same care and protection that you
do any rifle scope. Protect the lenses from the
elements as much as possible, and when they get
dusty, whisk away all of the loose dust with a
lens brush. Wiping dusty lenses only serves to
scratch them. Clean away fingerprints and water
spots by wiping the lens with a soft tissue damp-
ened with denatured alcohol.

With proper care, a pistol scope will give years
of satisfactory service. Not all handguns are can-
didates for a scope, but most of the specialty guns
deliver optimum accuracy only when scoped, and
a scope can be advantageous on many hunting
revolvers and semiautomatics. The secret to pistol
scope satisfaction, though, is to choose the right
scope for the shooting you do, then mount it
properly on your handgun.

*Conetrol mount in this photo
is typical of base and ring style
popular with handgun hunters. It holds the scope solidly
in place and when epoxyed
becomes a permanent fixture
of the pistol.*

ANNUAL UPDATE
Gun Developments

Jim Carmichel

According to what firearms manufacturers tell me, you haven't been buying many guns during the past few years. I can't say that I blame you, and I may as well confess that I haven't bought my share either. Every time I charge into a gunshop, money in hand, determined to buy something, I seem to be stricken with an attack of *dèjá vu*, because all the guns look

This article first appeared in Outdoor Life

like something I've owned before and don't necessarily want to own again.

The gunmakers like to cry on each other's shoulders and blame their lean market on the economy, but I'm not convinced that a shortage of dollars is the whole problem. As gun buyers, you and I are more selective than we have ever been. We want to be *excited* by guns that represent our thoughts on how a gun should look and feel and work. And if you are like me, you are turned off by gadgetry that adds nothing to function and bizarre designs that are different just for the sake of being different. I've said it before and I'll say it again: A gun doesn't have to be different to be good. If it is good that will be different enough.

The grand debut for the year's new guns takes place at a big splash called the SHOT Show (Shooting, Hunting, Outdoor Trade Show). That's where dealers buy their hardware and accessories for the coming hunting season and spies like me snoop around trying to discover what's new. The past couple of SHOT Shows have been fairly uplifting but in the final analysis were like a Chinese dinner—lots of interesting tidbits but nothing to stick to a man's ribs. This year the wares were a lot more exciting. There were guns I really *wanted!* Even my wife, who buys a gun about as often as I buy a castle in Siberia, took

one look at a 28-gauge Parker reproduction and bought it on the spot.

During the past few years, too many gunmakers had succumbed to the notion that they had to gutter fight for each other's market share. It appears that most manufacturers now realize that the surest way to stay in business is to stop playing front-office guessing games and simply do what they do best. The bitter lesson some had to learn is that no successful gun has ever been designed or made in the front office.

This year we find that gunmakers are indeed doing what each does best and showing lots of determination to do it even better. There are too many new guns to describe fully, but here's an idea of what you're going to find in your dealer's rack. Stop by and ask for more details.

REMINGTON

Remington earns center stage this season by virtue of a new bolt-action rifle and, at long last, a screw-in choke system for their shotguns. Their new bolt gun, called the Mountain Rifle, is built around a tried and true Model 700 action but is a half pound lighter than the M700 and at least ten times better looking. The stock styling is classic with a high, straight comb that makes the rifle handle like a custom-fitted shotgun. The overall appearance suffers slightly because of a rather bizarre checkering pattern on the grip panels. The design seems to have been originated by overzealous use of a french curve without enough consideration for classic restraint.

Available calibers are .280 Remington, .270 Winchester, and .30 06. Happily, Remington has also introduced a hot new .280 load with a 140-grain bullet. Team the new Mountain Rifle in .280 chambering with their new 140-grain bullet, and you've got one of the most effective and best-looking hunting combinations ever conceived.

I still don't understand why it took so long, but Remington has at last introduced their version of

Top to bottom: Winchester (Olin) Model 23 Classic side-by-side shotgun, now in 28-gauge and .410; Winchester (Olin) 28-gauge Parker reproduction; Remington Mountain Rifle, based on the Model 700 action; Browning reproduction of the Winchester Model 1886 lever-action rifle; Ruger Super Redhawk stainless-steel revolver with Ruger rings for scope.

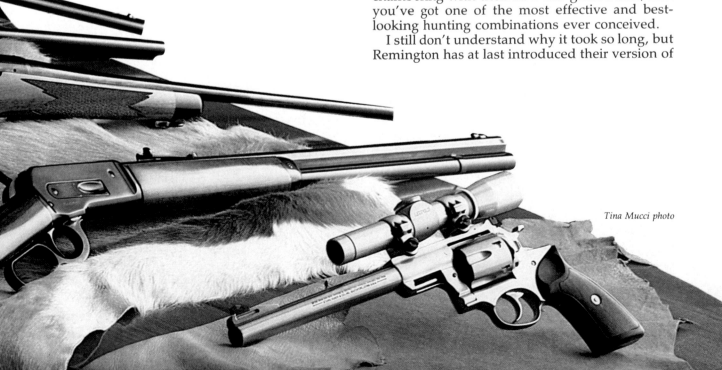

Tina Mucci photo

a screw-in choke system called, predictably, the "REM" Choke. Remington's engineers say they weren't dragging their feet but that there's a lot more to adapting a barrel to screw-in chokes than meets the eye. The choke tubes fit completely inside the muzzle and are removed with a key wrench. The Model 1100 autoloader, 870 pump and Sportsman Series can all be had with REM chokes in various barrel lengths but only in 12 gauge at present. When you heft one of the REM Choke-equipped shotguns you'll immediately notice a heavier, weight-forward feel. This is because there is a lot more steel in the barrels, which is said to be necessary. I expect that the folks at Remington find this additional mass of metal necessary because of possible liability suits. However, I expect that sales of REM Choke shotguns will not be what they might simply because of the additional weight. You'll see what I mean when you heft one of these new guns.

Remington's venerable 870 pump gun has been given a much-needed face-lift in the form of a dressed-up stock with a pattern of real cut checkering. This is a great shotgun, made even better.

A couple of old bolt-action favorites brought back by popular demand are Rimfire Rifle models 581-S and 541-T. It's always nice when a major manufacturer heeds customers' wishes, and these rifles have even more features than before. Possibly even Remington's great Model 3200 over/under shotgun can be made available again. After all, it is the gun that won the 1984 Olympic skeet event and deserves another chance. If it were dressed and trimmed down it would be one of the world's great shotguns.

Remington's custom shop is becoming a major operation. In addition to their fancy engraving and custom stock work, the shop is also turning out such pretty items as the Model 40-XR, a classy rimfire sporter built on the ultra accurate 40-X action. They are also making special-order rifles based on a left-hand version of the *short* M700 action. One custom shop product that is bound to grab everyone's attention is the Model 700 "KS" rifle. This is a space-age rifle with a synthetic stock reinforced with Kevlar. Weight is 6½ pounds, and most standard hunting calibers are available. Sooner or later this rifle will become so popular that it will become a standard production item.

WINCHESTER

By now most shooters know that there are two Winchesters. One is the Winchester of yore that manufactures the famous Winchester-brand ammunition and offers a classy line of shotguns. This Winchester has been a part of the Olin Corporation for decades and actually owns the world renowned trademark.

One of their guns, the neat and trim Model 23 side-by-side double, has a well-deserved reputation for dependable performance. Since its introduction in 1979, the M23 has been considerably upgraded by such refinements as screw-in Winchokes and optional English-style stocking, but the M23 has been available only in 12 or 20 gauge. This failing has been corrected in grand style by the addition of irresistible little M23s in both 28 and .410 gauge. The beauty of those two sub-bore doubles is that they are built on a scaled-down version of the M23 receiver. This gives them a grace and proportion of line not possible when small barrels are added to full-size frames. The two small bores are part of a four-gun "Classic" series. Model 23 Classics in all four gauges have 26-inch barrels with ventilated ribs and come in Improved Cylinder and Modified choking.

The other "Winchester," which uses the name under license from Olin, is U. S. Repeating Arms Company. They build rifles and shotguns in the original Winchester plant in New Haven, Connecticut, and have continued such legends as the Model 94 lever rifle, Model 70 bolt rifle and the high ticket M21 double. This year, like Remington, they are bowing to the future and offering the Model 70 with a fiberglass stock. But unlike Remington, which makes fiberglass-stocked rifles more or less one at a time in their custom shop, Winchester offers their Model 70 "Winlite" as a standard production item. This means that the price is not all that much more than you'd pay for a similar rifle with a walnut stock. Winchester gave a lot of thought to their M70 Winlite and added such technological features as Kevlar and graphite along with the usual epoxy and fiberglass. This means the stock is lighter and stronger than plain fiberglass. Weights are 6½ pounds for the Featherweight and 6¾ pounds for the Sporter rifles. This is about a pound lighter than the same models with wood stocks. Of course the walnut stocks are still available and will be for a long time.

If you're a dedicated fan of European calibers you'll probably dance a polka when you hear that the Model 70 Featherweight is being offered in 6.5×55mm. Swedish chambering. This is one of those sweet little rounds that hardly kicks at all but drills through brush like a truck and drops deer in their tracks.

WEATHERBY

Weatherby was one of the first major gunmakers to offer synthetic stocks as an alternative to traditional wood. Their Fibermark stocks, according to Roy Weatherby, have proven more popular than Weatherby predicted. I tried one a few years ago and was much impressed with the rifle's accuracy. This year, however, Weatherby is taking a step back in time and offering—now get this—a Weatherby Mark-V rifle without the characteristic bright glossy finish! It is called the Euromark and the walnut stock features a hand-rubbed, satin oil finish. Likewise the metal has a soft, nonglare blueing.

RUGER

Ruger is another major manufacturer offering synthetic stocks. A new version of their bolt rifle is the M77RL which features a stock of resin-impregnated laminates.

The stock is nearly indestructible, totally warp-free and so waterproof that you can clean it in the dishwasher. Inside rumors have it that Ruger is also at work on a fiberglass stock for their Model 77 rifle.

Have you heard of ergonomics? It's the science of contouring a utensil so it not only feels comfortable in the user's hand but also generates the least possible discomfort and fatigue. Ruger's new .357 Magnum revolver, the GP-100, is an ergonomic as well as technological masterpiece. It includes too many features to describe here but when you hold it in your hand you'll realize that Ruger isn't kidding when it says that the GP-100 represents a whole new generation in revolver design.

Another typically Ruger innovation is the new Super Redhawk revolver which is specifically designed for scope use. The frame has been lengthened and is notched for Ruger's scope rings. This means that scope mounting is quick, easy and strong.

MOSSBERG

The problem with describing Mossberg's line of guns is that there are so many of them. Basically Mossberg offers four groups of firearms: (1) A line of imported bolt-action sporting rifles, (2) some truly excellent Japanese-made autoloading and pump-action shotguns, (3) a line of U.S.-made pump and autoloading shotguns, and (4) U.S.-made shotguns for police and self-defense uses.

Mossberg's U. S. plant is now wholly involved in the production of shotguns. They feel that this is what they do best, and they intend to sell more shotguns than anyone else. The backbone of their domestic line are the Model 712 autoloaders and Model 500 pump actions. The M712, which comes in 12 gauge only, will handle *any* 12-gauge shell from the light 2¾-inch load to the heaviest 3-inch

magnum. You don't have to make any adjustments for different loads—just load and pull the trigger. The Model 500 pump guns, like their autoloaders, come in a tremendous variety of variations in barrel lengths and chokes and either 12 or 20 gauge. Check a Mossberg catalog to see what's available. When you check the prices, you'll discover you can still get a lot of shotgun for a moderate amount of money.

I've tried most of the Mossberg shotguns on both waterfowl and upland game, and either they are mighty easy to hit with or I was unusually lucky. The stock dimensions, which are straighter and fuller than most, are a definite aid. If you want to spend a bit more money, the Mossberg Regal Series offers real walnut, cut checkering, and bright blueing.

The Series 1000 autoloading shotguns need little introduction because they have already earned a reputation as one of the best gas-operated autoloaders ever made. The M1000 Mossberg Waterfowler is my pick as the best duck gun available. It has it all, quality, looks, and performance.

MANNLICHER

Have you been wondering whatever became of Mannlicher? These sleek-looking and slick-working rifles were once staple items in high-class gunshops, but over the years they more-or-less faded away. The problem was not with the Austrian-made rifles, because in Europe they are as popular as ever. U. S. importers just didn't seem to have their hearts in selling the trim sporters. Now, that has all changed, and the House of Mannlicher is Gun South Inc. The people at Gun South know guns, and they are determined to make Mannlicher the household word it once was. I'm glad an old favorite is around again.

MARLIN

"Papoose" is an apt name for Marlin's little take-down-tuck-away .22 Rimfire. This seven-shot, clip-fed self-loader takes down in a couple of seconds to store in a neat padded case. It's a handy item on camping, canoeing, or 4×4 excursions, and the case floats. The Papoose comes complete with the case and a 4× scope. It's a clever package, and I intend to buy one for the survival kit in my Cessna.

BROWNING

Browning's sweet little thing for this year is the bolt-action .22 Rimfire Rifle that is almost a dead ringer for their Centerfire A-Bolt rifles. It has the same feel and same looks. And it is only 1½ inches shorter than the centerfire version. In fact, some of the parts are interchangeable between the centerfire and rimfire versions of the A-bolt rifles. Also virtually identical to the Centerfire's version is the Rimfire's quick, 60-degree bolt lift and gold trigger which is adjustable for let-off. The finish and checkering show lots of class and the rifle comes with two clips, one of five-shot capacity and the other a 15-shot. The receiver is grooved for rimfire-style scope rings and also drilled and tapped if you want to mount more sophisticated-looking scope rings. Iron sights are available if you want them. Otherwise the barrel comes slick and clean for a more streamlined profile when using a scope.

Browning has always made a stylish rimfire rifle, going back to the original John Browning autoloader design. This new A-Bolt Rimfire fits nicely into the modern trend towards bolt-action rimfires with grown-up proportions and stylish good looks.

New this year, Browning is also offering a shortened and lightened version of their gas-operated B-80 autoloading shotgun. Called the Upland Special, and available in both 12 and 20 gauge, the Upland Special features a 22-inch barrel and a straight, English-style stock. The weight is about a half pound less than that of their standard shotguns. The Browning Upland Special comes with Invector choke tubes so that you have a light, fast-handling, all-purpose shotgun.

EXEL

Exel, you know, is the firm that is importing a line of good-looking but priced-right line of European-made shotguns. I've been particularly excited about their small-gauge doubles and will have more to say on them in the future. Side-by-side doubles are making a comeback, and Exel has the right guns at the right time. For their catalog, use the address at the end of this article.

CAMEX-BLASER

This is the West German firm that last year introduced the "Ultimate" rifle: a lightweight bolt action of radical design that easily converts from one caliber to another. Now Camex-Blaser has added a trim single-shot rifle and an air pistol attachment that is both novel and useful. It is a CO_2-powered unit that converts your Colt autoloading pistol (1911-style) to fire .17 caliber pellets. This is a wonderful way of practicing with your Colt pistol because you have the same trigger pull and general feel as when firing live ammo.

THOMPSON-CENTER

Here's a firm that can never be overestimated, either in the quality of their products or in the soundness of their ideas. They know what they're doing, and they do it extremely well. The latest addition to their Contender "Shooting System" is a rifle-style carbine stock on the basic break-open receiver. Add an interchangeable carbine barrel in any of six calibers .22 RF to .30-30 and you have a compact little rifle. Or, get this, add the .410 barrel, and you've got a shotgun with ventilated rib. This dynamic firm has lots of other new items to tell you about.

IVER JOHNSON

Space-age technology has come to this venerable gunmaker in the form of a single-shot, rimfire youth-size rifle that features a rock-bottom price and nigh-on-to-indestructible stock of high tech material. The basic design is simple and about as foolproof as a gun can be. That's a blessing for parents considering a first gun for the kids. This one should last several generations.

THE SWEETHEARTS FOR 1986

Here are five gems for this year—the guns I won't be able to do without. They are the Alpha Bolt Rifle, Browning's reproduction of the 1886 lever rifle, Heckler & Koch's super accurate BASR Rifle, Kimber's latest rifles and the 28-gauge Parker Reproduction by Winchester.

The Alpha Bolt Rifle isn't new, but every year it gets better. What started out as a solid, but plain, lightweight centerfire hunting rifle has evolved into a stylish and sophisticated piece of hunting equipment. This year a long-action version of the Alpha will be introduced that will handle such big magnum rounds as the .357 H&H. This will make the Alpha a truly world-class hunting rifle.

The Winchester Model 1886 lever-action rifle was quite simply and unmistakably the greatest lever-action rifle ever manufactured. It was a big rifle for big cartridges (such as the .50-110 Winchester) built by an inspired band of craftsmen. The rifle was created by no less a designer than John M. Browning, so what could be more fitting in the rifle's 100th birthyear than for it to be reproduced by Browning Arms?

The sample I saw certainly did justice to the original in this one-time-only offering.

Heckler & Koch, as everyone knows, is a West German gunmaker specializing in high tech handguns, military hardware and some of the best-made autoloading sporting rifles available.

Their introduction of a U.S.-built bolt-action ultra-accurate rifle therefore comes as something of a shock, but it is a pleasant shock. Basically for varmints and designed and built from scratch, the rifle features all the accuracy technology available and then some. A five-shot test group fired at 100 meters (110 yards) with Federal factory .308 ammo measured .220 inches. That's right, less than a quarter-inch between the centers of the widest shots! Think of how well your handloads will shoot. Several calibers are available from .22–250 to .300 Winchester Magnum.

Kimber of Oregon flew in the face of logic a few years ago with a stylish rimfire that most experts said wouldn't sell. Well, they have sold very well indeed, which is a credit to the good taste of American gun buyers. Each year Kimber offers something nicer than before, such as this year's Mannlicher-stocked Rimfire complete with butter-knife bolt handle. But what they have that will really turn you on is a new centerfire that harks back to the great prewar Mausers and the pre-1964 Model 70 Winchester, plus some nice new touches. Even custom rifle builders will be hard-pressed to match this one.

I won't have to dig into my paltry savings account for this one because my wife beat me to it. It's a 28-gauge version of the Parker Reproduction being built in Winchester's Japanese plant. When Tom Skeuse, who manhandled the Parker project into being, introduced the sweet 20-gauge version a few years ago, I knew he had a winner. These are beautifully made and faithful copies of the great Parker double and probably the finest shotgun ever made in Japan. The 28-gauge version, new this year, is built on a smaller frame and is utterly irresistible. Stock dimensions are just right, down to the skeleton buttplate, and you have a choice of straight or pistol grips, beavertail or slender forearm, and single selective or double triggers. Selective ejectors are standard and barrels are 26-inch or 28-inch. Weight is a few ounces over five pounds and the gun comes in a fitted leather case.

Well, there's some, but not all, of the new guns you'll be seeing. For a first-hand look, visit a dealer. Or for more details, write the makers. Here's where to write:

Alpha Arms, Inc.
12923 Valley Branch
Dallas, TX 75234

(Anschutz)
Precision Sales International PSI
P. O. Box 1776
Westfield, MA 01086

Browning Arms
Route 1
Morgan, UT 84050

Camex-Blaser USA Inc.
308 Leisure Lane
Victoria, TX 77904

EXEL Arms of America
14 Main Street
Gardner, MA 01440

(H&K)
Heckler & Koch, Inc.
14601 Lee Road
Chantilly, VA 22021

Iver Johnson
2202 Redmond Road
Jacksonville, AR 72076

Kimber of Oregon, Inc.
9039 S.E. Jannsen Road
Clackamas, OR 97015

O. F. Mossberg & Sons, Inc.
7 Grasso Street
North Haven, CT 06473

(Mannlicher)
Gun South, Inc.
P. O. Box 129
108 Morrow Avenue
Trussville, AL 35173

Marlin Firearms Co.
100 Kenna Drive
New Haven, CT 06473

Parker Reproduction Division
Reagent Chemical & Research, Inc.
17th & South Hall Street
Webb City, MO 64870

Remington Arms Co., Inc.
939 Barnum Avenue
P. O. Box No. 1939
Bridgeport, CT 06601

(Ruger)
Sturm, Ruger & Co.
Southport, CT 06490

Thompson-Center Arms
P. O. Box 2426
Rochester, NH 13867

U. S. Repeating Arms Co.
P. O. Box 30-300
New Haven, CT 06411

Weatherby, Inc.
2781 E. Firestone Blvd.
South Gate, CA 90280

Winchester-Olin Corp.
120 Long Ridge Road
Stamford, CT 06904

Index

Aberdeen Proving Ground, 37
Accuracy
  checking, importance of, when
    changing primers, 117
  handgun, enhancing, 159–63
  life, of stainless steel barrels, 149
  of rifle barrels, 141
Adolph, Fred, 102
African Game Trails, by Theodore
    Roosevelt, 85
African Rifles and Cartridges, by John
    Taylor, 167
Akin, Al, 25
Alley, Dean, Supply Co., 8
Alpha Arms, Inc., 182
America's Munitions 1917–1918, by
    Benedict Crowell, 36
American Rifleman, 38, 100, 103, 157
Ammunition
  black-powder match, 93–94
  See also Cartridges
Anschutz (Precision Sales Int'l.—
    PSI), 90, 93
Aperture sights
  advantages and disadvantages of,
    164–66
  assembly, requirements for, 166

Ballistic Products, Inc., 139
Ballistics
  handgun cartridge, 43(*table*)
  of hunting cartridges, 118–22
  Rifle Cartridges, Point-Blank
    Trajectories and Remaining
    Energy Levels of,
    120–21(*chart*)
Barrelmakers, 141, 148
  Baiar, Jim, 20
Barrels
  lapping, 148–49
  manufacture of, 142–45
  rifle, accuracy of, 141
  rifling methods compared,
    145–48, 149(*chart*)

stainless steel, accuracy life of, 149
  See also Barrelmakers
Bausch & Lomb Inc. *See* Binoculars,
    Scopes
Beecham, Greg, 19
Beecham, Tom, 19
Benchrests, use of, in sighting-in,
    8–9
Berdan, Hiram, Colonel, 110
Beretta U.S.A., handguns, military
    testing and use of, 39
Bergmann
  handguns, military testing and
    use of, 35
Bianchi Int'l., Inc., 171
Biesen, Al, 5
Big Game Rifle, The, by Jack
    O'Connor, 169
Billingsly, Ross, 20
Binoculars
  features of, 24–25
  light-gathering ability of, factors
    determining, 23–24
  manufacturers of, 25, 26
Bishop, E.C., & Son Inc., 33
Black-powder target shooting,
    87–94
  ammunition, 93–94
  DCRA rules for, 87–89, 94
  gunner's quadrants, 93
  rifles, 89, 92–93
Bolt Action Rifles, by Frank de Haas,
    103
Bonanza (Forster Products Inc.), 116
Bonillas, Dan, 65–71
Boxer, Edward M., 112
Branham, Burl, 65–71
Briarbank Ballistic Laboratory, 122
Browning Arms, 181–82
  handguns, military testing and
    use of, 39
  shotguns, 73
Bullet makers, custom, Bill
    McBride, 77

Bullets
  for big game, 128–29
  for black-powder shooting, 94
  Newton, 103(*illus.*)
  for pronghorn hunting, 29–30
Bushnell Optical Co. *See* Scopes

Cable, Lee, 19
Camex-Blaser USA Inc., 182
Cartridges
  factory, consistency of, 6
  handgun, 42–45
  Newton, 101–02, 104
  for pronghorn hunting,
    28–30(*illus.*)
  rifle, compared, 118–22
  .280, 95–97
  .220 Weatherby Rocket, 106
  Weatherby Magnums, 107,
    108–09
  for Winchester Model 70, 11, 13,
    16–17(*illus.*)
  See also Ballistics
Cartridges of the World, 99
Case sizing, full-length vs. neck,
    129–30
Cavey, Bruce, 26
Centaur Systems, Inc., 160
Central sights, 90, 93
Charter Arms Corp., 40
Collimators, manufacturers of, 8
Colt Firearms, 40
  handguns, military testing and
    use of, 34–39
Corbin, Dave, 76
Crossman, E.C., Captain, 97

Deutsche Waffen und Munitionsfabriken
    (DWM), 34
Dixie Gun Works, Inc., 170
Dominion of Canada Rifle
    Association (DCRA), 94
  rules for long-range black-powder
    target shooting, 87–89

Douglas Barrels, Inc., 141
Duplex. *See* Reticles

Engravers
   Dowtin, Bill, 20
   Marktl, Franz, 19, 20
Etchen, Rudy, 64–71
EXEL Arms of America, 182

Fabrique Nationale
   handguns, military testing and
      use of, 39
Fajen, Reinhart, Inc., 4
Forsyth, Alexander, Reverend, 110
Foster, Karl, 75
Freedom Arms (L.A.R. Mfg. Co.),
      40, 42

Gun Control Act and handgun
      specifications, 57
Gun Digest, 44
Gunmakers, Ulrich Vosmek, 153
Gunner's quadrants, 93
Gunshot Injuries, by Col. Louis
      LaGarde, 34

Handbook for Shooters & Reloaders, by
      P. O. Ackley, 118
Handgun manufacturers and
      importers
   Ranger Mfg. Co., Inc., 56
   Stoeger Industries, 56
   Thalson, 57
Handguns
   accuracy of, enhancing, 159–63
   automatics, 43–44
   for big game, 40–45
   bolt-actions, 44
   cartridges, ballistics, 43(*table*)
   cartridges, for big game, 44–45
   Colt .45, 34–39
   for combat competition, 159
   Mauser C/96 Broomhandle, 46–55
   military testing and use of, 34–39
   Novak sights for, 162–63
   Quadra-Lok system, 160(*illus.*)–62
   revolvers, 40–43
   scope power, recommended, for
      hunting with, 175(*table*)
   scopes for, 172–77
   Thompson-Center Contender, 44
   Walther PPK, 56–59
Handloader, 20
Handloading
   for big-game hunting, 128–34
   equipment and manufacturers,
      131, 134–39
   load choice, accuracy vs. velocity,
      133
   load performance, testing, 132–33
   at the range, 130–31
   for shotguns, 123–27
   steel shot, 134–40
Hart, Atkinson (H-S Precision Inc.),
      141

Heckler & Koch, Inc., 182
   handguns, military testing and
      use of, 39
Holsters, 56
Huey, Marvin, 20
Hydro Press, 76

Interarms Ltd., 40
Ithaca Gun Co., shotguns, 72, 73
Iver Johnson, 183

Keith, Elmer, 166, 169
Kimber of Oregon, Inc., 182
Knoble handguns, military testing
      and use of, 35
Koshollek, Emil, 157
Kuhn, Bob, 20

Lawrence, George, Co., The, 170
Lee Precision, Inc., 116
Leupold & Stevens Inc. *See* Scopes
Luger handguns, miltary testing
      and use of, 34, 35
Luger, Georg, 34

McMillan Rifle Barrels U.S. Int'l.,
      141
Magnum Sales, Ltd., 171
Mannlicher. *See* Steyr-Daimler-Puch
Manurhin (Matra-Manurhin Int'l.,
      Inc.), 56
Marathon Products Inc., 4
Marksmanship, by Gary Anderson,
      168–69
Marlin Firearms Co., 181
Marsh, Mike, 20
MATCO, Inc., 141
Mauldin, Bill (cartoons), 37
Mauser C/96 Broomhandle, 46–55
   chronology of, 52(*chart*)
Mauser, Peter Paul, 46
Mayes, Wayne, 64–71
Mayville Engineering Co., Inc.
      (MEC), 134
Michaels of Oregon, 170
*Military Discipline—The Complete
      Gunner*, by Thomas Venn, 167
Miller, David, Company, 5
Mossberg, O. F., & Son, Inc., 181

Newton, Charles, 101–04
*Newton, Charles, Father of High
      Velocity*, by Bruce M.
      Jennings, Jr., 101
Nikon Inc. *See* Binoculars, Scopes

Obermeyer, 141
Ohye, Kay, 65–71
*100 Years of Shooters and Gunmakers
      of Single Shot Rifles*, by Gerald
      Kelver, 89
Optical equipment, 22–24

care of, 26
terms, glossary of, 22

Pachmayr Gun Works, 170
Parker-Hale sights, 90, 93
Parson, Leon, 20
Pennsylvania Arms Co., 73
Penrod, Mark, 20
Penrod Precision, 20
Pistolsmiths: Wayne Novak, 160,
      162–63
Powder
   for big-game loads, 129
   smokeless, 112
Power Plus Enterprises, 76
Practical Marksmanship, by Capt.
      Melvin M. Johnson, USMC
      (Ret.), 169
*Practice and Observations with Rifle
      Guns*, by Ezekiel Baker, 168
Prall, Jimmy, 65–71
Primers, 110–17
   benchrest, 116
   for big-game loads, 129
   designations, 117(*table*)
   magnum, 116
   seating, 113–16
   shotgun, 117
   storage of, 117
Priming systems, 116
Purdey, 2

Ranger Mfg. Co., Inc., 56
RCBS, Inc., 116
Redfield Gun Sight Co., 93
Reinders, Vic, 64–71
*Reloaders' Guide for Hercules Smokeless
      Powders*, 123
Reloading. *See* Handloading
Remington Arms Co., 179–80
   handguns, 44
   shotguns, 73
Restocking a Rifle, by Alvin Linden,
      157
Reticles, 26
Rifles
   black-powder match, 89, 92–93
   bolt-action, 30, 33
   .450 Holland & Holland, 86
   lever-action, 31
   Newton, 102–04
   percussion, early, 110
   for pronghorn hunting, 30–31, 33
   .450 Rigby, 86
   .280 Ross, 95–100
   Ruger No. 1, 19–20
   semiautomatic, 31
   single-shot, 30–31
   slide, 31
   1903 Springfield .30–06, 85
   Weatherby, 107–08
   Winchester
      Model 70, 10–17
      Model 95, 85–86
   See also Rifles, custom

Rifles, custom
    features of, 4–5
    kits, 4
    makers, 2, 4, 5, 102, 150–57
    prices, 4–5
The Rifle, 20
Rifling
    button, 146–47
    cut, 145–46
    Electrochemical (ECR), 149
    methods compared, 149(*chart*)
    rotary forging, 147–48
Roosevelt, Theodore, 83–86
Ross, Charles, Sir, 98–100
Ruger, J. Thompson, 20
Ruger, Sturm, & Co., Inc., 19–20,
        40, 181
    No. 1, 19
    21 North Americans, 19–20
Ruger, William B., 19, 20

Savage Industries, Inc. handguns,
        military testing and use of, 35
Scherer, Ed, 64–71
Scopes
    features of, 25–26
    fixed- vs. variable-power, 31
    handgun, features of, 172–74
    light-gathering ability of, factors
        determining, 23–24
    manufacturers of, 25, 26, 31, 33
    mounting, on handguns, 176–77
    mounting systems, for handguns,
        174–76
    for pronghorn hunting, 31–33
    spotting, 26
SGW, Inc., 141
Sharps Rifle Company, 89
Shaw, Joshua, 110
Shilen Rifles, Inc., 141
Shooter's Bible, 44
Shot
    chilled vs. hard, 80
    See also Steel shot
Shotgunning tips, 64–71
Shotguns
    fit, 60–63
    patterning, 78–82
    wads, weights compared,
        127(*table*)
    See also Slug guns, Slug wads,

choice of, 123–26; patterns
        and, 126–27
Shots Fired in Anger, by John
        George, 168
SIG (Sigarms, Inc.) handguns,
        military testing and use of, 39
Sighting-in, 6–9
Sights
    aperture, advantages and
        disadvantages of, 164–66
    iron, 90
    micrometer peep, 90
    See also Scopes
Simmons Outdoor Corp. *See* Scopes
Sling'N Things, 170
Slings
    evolution of, 167–71
    manufacturers and suppliers of,
        170–71
    shooting with, 168–69
    styles of, 169, 170
    swivels, 169–70
Slug guns, 72–74
    competition, 72
    manufacturers of, 72–73
    shooters' organizations, 77
Slugs, 74(*illus.*)–76
    Brenneke, 75
    BRI Sabot 500, 76
    Foster, 75
    swaged, 76–77
    Vitt-Boos Aero Dynamic, 76
Smith & Wesson, Inc., 40
    handguns, military testing and
        use of, 35–39
    shotguns, 73
SSK Industries, Inc., 45
Star handguns, military testing and
        use of, 39
Starrett, Dave, 64–71
Steel shot
    hulls, 140
    press conversion kits, 134–36
    reloading, 134–40
    scales for, 137–38
    size of, 140
    wads, 139
    wad-slitting tools for reloading,
        137
Steyr-Daimler-Puch, Gun South,
        Inc., handguns, military

testing and use of, 30
Mannlicher rifles, 181
Stockmakers
    Golueke, Charles, 158
    Linden, Alvin, 150–57
    Mews, Leonard, 108, 158
    Pohl, Henry, 33
    Shelhamer, Thomas, 158
Stock manufacturers:
    Bishop, 33
    Fajen, Reinhart, 4
Stoeger Industries, 56
Swanson, Gary, 19
Swivels. *See* Slings

Tasco, 8
Taylor, John "Pondoro", 166
Telescopic sights. *See* Scopes
Thalson importers, 57
Thompson-Center Arms, 182
    Contender, 44

U-Load, Inc., 137, 139
United States Repeating Arms, 17

Waffenfabrik, Carl Walther, 56
Walther
    handguns, military testing and
        use of, 39
    PPK, 56–59
        manufacturing and importing
            of, 56–57
        specifications, 58(*chart*)
Weatherby, Inc., 180–81
Weatherby, Roy, 101, 105–09
Wesson, Dan, Arms, 40, 42
Whelen, Townsend, Colonel, 97, 99,
        152
White-Merrill handguns, military
        testing and use of, 35
Williams Gun Sight Co., 170
Winchester, 180, 182
    Model 54, 10
    Model 70, 10–17
    Model 95, 85–86
Wolfe, Dave, 20
Woytek, Ken, 24
Wundhammer, Ludwig, 152

Zeiss, Carl, Inc. *See* Binoculars,
        Scopes